Soil Mechanics—New Horizons

Contributors

O. G. Ingles B.A., M.Sc. (Tas.), F.R.I.C., A.M.Inst.F., C.Eng.
Principal Research Scientist, Division of Applied Geomechanics, C.S.I.R.O., Australia.

P. Lumb M.Sc., A.M.I.C.E.
Reader in Civil Engineering, University of Hong Kong

B. G. Richards B.E. (Hons.), M.S.
Principal Research Scientist, Division of Applied Geomechanics, C.S.I.R.O., Australia.

I. K. Lee B.C.E., M.Eng.Sc., Ph.D., F.I.E. (Aust.), M.A.S.C.E.
Professor of Civil Engineering, Head of Department of Civil Engineering Materials, The University of New South Wales, Australia.

S. Valliappan B.E. (Annam.), M.S. (Northeastern), Ph.D. (Wales), A.M.I.E. (India), M.A.S.C.E.
Senior Lecturer in Civil Engineering, Department of Civil Engineering Materials, The University of New South Wales, Australia.

J. R. Herington B.E., M.Eng.Sc., M.I.E. (Aust.)
Staff Engineer, Dames and Moore, Sydney, Australia.

H. G. Poulos B.E., Ph.D., M.I.E. (Aust.), M.A.S.C.E.
Reader in Civil Engineering, University of Sydney, Australia.

Soil Mechanics— New Horizons

Edited by
I. K. LEE

B.C.E., M.Eng.Sc., Ph.D., F.I.E. (Aust.), M.ASCE.

Professor of Civil Engineering,
Head, Department of Civil Engineering Materials,
The University of New South Wales, Australia.

AMERICAN ELSEVIER PUBLISHING COMPANY, INC.

NEW YORK

Published in the United States by
American Elsevier Publishing Company Inc.,
52, Vanderbilt Avenue,
New York, N.Y. 10017

ISBN 0-444-19542-4

Library of Congress Catalog Card No.: 73-17948

First published in 1974

Filmset by Thomson Litho Ltd, East Kilbride, Scotland
Printed in England by Fletcher & Son Ltd, Norwich
Bound in England by Richard Clay (The Chaucer Press) Ltd, Bungay, Suffolk

Preface

•

Since the publication of Soil Mechanics—Selected Topics in 1968 the authors have been encouraged to write a second volume presenting new developments in topics previously discussed, and to extend the scope by including certain fields not considered in the earlier book. Thus the present book was written as a complementary volume by updating certain aspects previously discussed, and the opportunity has been taken to extend the range of topics.

The subject matter consists of four chapters on material behaviour and properties and three chapters on analysis of soil structures. This follows the general theme previously established except that a separate discussion on experimental and field techniques is not included.

Chapter 1 (Ingles) is an in-depth discussion of the techniques of compaction, and, in effect, traces the development of the multiplicity of compaction techniques now available to the geotechnical engineer.

Chapter 2 (Ingles) extends the earlier (1968) discussion by summarising the state of art of stabilisation. The chapter includes a discussion of the selection and design criteria, and the statistical control techniques now available. This statistical approach leads to the more detailed chapter on application of statistics in soil mechanics (Lumb), and the developing significance of this tool is made evident in many areas of geotechnical engineering.

Developments in the general area of the behaviour of unsaturated soils have been extensive but the practical application of this knowledge to engineering situations poses certain difficulties. Chapter 4 (Richards) not only gives an account of the physics of unsaturated soils but also shows the application of an analytical technique to the prediction of the settlement deformations of structures supported by or composed of a soil in an unsaturated state. This chapter extends the 1968 discussion (Chapters 2 and 3).

Chapter 5 (Lee and Valliappan) extends the 1968 discussion (Chapter 3) on consolidation. Recent developments in the analysis of raft foundations is then considered in some detail, and the application of the finite element method to this problem, and to the settlement analyses considering non-linear behaviour, is included.

Certain aspects of retaining structure behaviour, such as the effect of the nature of the wall movement, can now be resolved by treating the retained soil as a perfectly plastic associated or non-associated flow rule material. Chapter 6 (Lee and Herington) extends the earlier discussion (Chapter 6) and details the techniques for establishing bound solutions, discusses the validity of traditional solutions, and illustrates the applicability of the plasticity theory.

The final chapter on the analysis of piles and pile groups (Poulos) has been written with the object of showing the application of the linear elastic analyses to pile systems. A series of comparisons between field and laboratory studies and the predicted behaviour shows that the analytical approach based on the simplified model of a soil is an extremely valuable predictive tool.

The editor deeply appreciates the ready cooperation of his colleagues in preparing the book and only hopes their work receives due acknowledgement. He would also like to thank his staff for their encouragement and to Mrs. B. Hegh and Miss J. Bain for their editorial assistance.

The authors would like to acknowledge the ready cooperation of authors, Institutions, Societies, publishers and editors who have granted permission for the reproduction of diagrams and tables from their publications. The editor is particularly grateful for the support given by the Australian Research Grants Committee for the work described in chapters 5 and 6.

Contents

Chapter 1

Compaction

O. G. Ingles

1.1   BASIC PRINCIPLES

Perhaps the single most important task in any earthwork is to ensure good compaction. Compaction substantially influences the future behaviour of any earthen structure inasmuch as poor compaction leads to poor strength, high permeability, large settlements and lower erosion resistance. The dangers of slip or collapse are enhanced; and for expansive clays, the potential swell or shrink is more rapidly realised.

It is often difficult to recompact a disturbed soil to equal or greater density than in its natural state, without the use of heavy mechanical equipment. It should be noted that weight, however, is not the only requirement. Indeed it is a common fallacy that heavy machinery will effect compaction according to its size and weight, whereas in fact the relevant factor is the *bearing pressure* exerted at the soil surface. One obvious instance of this principle is the poor compaction effect of an elephant, but the same principle applies equally to tractors, bulldozers and other equipment designed not to sink into soft ground. *The first principle of compaction is therefore to ensure a substantial contact pressure with the soil.*

The earliest forms of earthen construction, houses and roads, tacitly recognised a need for compaction, either by ramming (pisé bricks) or by trafficking the soil. One of the earliest references surviving is that of the Chinese engineer Chia Shan (178 B.C.) who wrote, in the Chih Yen, 'The road was made very thick and firm at the edge, and tamped with metal rammers.' It was, moreover, quickly apparent that there existed a common aid to the compaction of almost any dry soil, i.e. water. Dry clays could be more easily densified and formed into the required shapes for simple building bricks by puddling and drying; roads and paths became firmer under the repeated traffic of wheeled vehicles and animals (except in very wet conditions where no side drainage was provided) and fixed traderoads developed. In the earliest forms of earth dam construction, such means and aids to compaction were commonly practised, as for instance one of the earliest accounts of Australian farm practice, following, reveals.

To construct a three-ply dam. Three essentials you need are, the catchment, the correct type of clay in the subsoil[†] and suitable land for what you wish to grow.

1. The area for the dam and bank must have all the top soil removed.[‡]
2. The clay must be well ploughed,[§] with two or three horses. Then each horse to pull a small scoop.
3. The first tip to be made at the end of where the dam is to be, and to continue until the first layer is completed, as the scoop, and the horses feet, tramp the clay as it is tipped, which has a pugging[||] effect.
4. The second row is completed, then the third.
5. As the bank rises, then only two rows are laid, and finally only one layer is used. During the whole operation the tramping of the horses feet consolidate the dam.[¶]

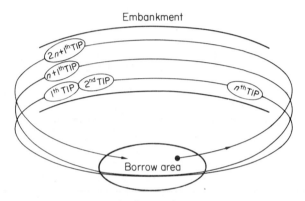

Figure 1.1   Construction sequence

In many respects, this account embodies principles already practised for more than 2000 years since Sunshu Ao, the Chinese hydraulic engineer, built the Peony Dam in Northern Anhui in 600 B.C.

Too much water was obviously undesirable, as the soil became too sticky, or even worse. Too little was equally undesirable; and clearly a certain level of moistness in each soil gave the easiest and best compaction. Thus, *the second principle of compaction is that the moisture content at which the soil is compacted determines the effectiveness of the contact pressure applied.* This moisture content is known as the optimum moisture content and for a given compactive effort is an unique property of each soil. The relationship between initial moisture content and the dry density of earthen material is of the form shown in *Figure 1.2*. Also shown on *Figure 1.2* is a line known as the zero air voids line, which represents the densities which might theoretically[††] be achieved if no air voids remained in the compacted mass. There are several important aspects of this figure.

[†] This is described as 'mottled yellowish' and no doubt refers to the majority of the Australian coastal podzols which have developed into a duplex profile.

[‡] 'About 0.3 m deep'. *This exposes clay which (in duplex soils) is normally close to optimum moisture content for compaction.*

[§] 'With a large single furrow plough.'

[||] A clear description of soil at the correct moisture content for optimum compaction with given compactive effort.

[¶] A reconstructed sketch of the procedure is shown in *Figure 1.1*

[††] The position of this line depends on the density of the soil solids. That shown is for 2.68 gm cm^{-3}, the usual average value for soils. Higher densities move it to the right, lower densities to the left, but always remaining approximately parallel to the line shown.

Firstly, it will be seen that to the right of the optimum moisture content, compacted density falls off *due to an inability to expel excess water* (even the air voids, although small, are not further diminished). This loss of compacted density, wet-of-optimum, may be ascribed to pore pressure in the water phase, which cannot dissipate. This is true in a confined situation; but where little significant confining force exists, the material simply flows away from the compaction zone due to loss of shear strength in the soil.

Secondly, it will be seen that to the left of the optimum moisture content, the compacted density falls off in such a way that air voids increase extremely rapidly. This loss of density may be attributed to either or both the factors of increasing shear resistance to the compaction force, or increasing resistance

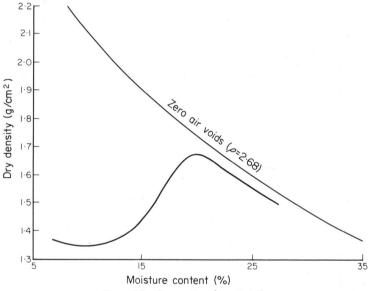

Figure 1.2   Compaction characteristics

to grain fracture (and hence an increased resistance to any densification due to the creation of a range of particle sizes capable of tighter (denser) packing than the original soil).

However it is not necessarily most advantageous to compact at the optimum moisture content determined from measurements of the dry density–moisture relationship. The primary consideration for clayey soils should always be the *equilibrium moisture* content of the structure in service, since a soil compacted wet or dry of this final equilibrium moisture content will either shrink or swell after construction with deleterious effect on the work. Thus, for clayey soils, it is better to compact at the maximum load which the equilibrium moisture content of the soil will sustain. In arid regions, this calls for much heavier rollers than in wetter regions. Moreover, whenever the moisture content of a clayey soil requires adjustment, it must be remembered that *different* compaction effects arise according to whether the soil is being wetted-up or dried out. This is due to the very slow redistribution of moisture within a dense clay, a process which may take many days or even weeks. Thus wetting-up tends to aid compactability by creating low friction skins on the clay lumps, and drying out to oppose compaction by creating higher friction skins on the clay lumps. This is true both in field and in laboratory work, being a source of much error in the latter case.

In setting requirements for compaction based on moisture content and density, it is also important to keep in mind the *purpose* of the work. Thus dams (which require low permeability) should be constructed slightly wet-of-optimum, whereas stabilised pavements, especially lime stabilised clays, are best constructed slightly dry-of-optimum. The effects of common stabilisers on soil compaction are sufficiently important to warrant specific statement. In the case of cement stabilised soils it has been found that higher strengths

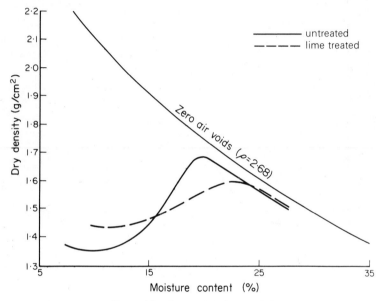

Figure 1.3   Compaction characteristics

are obtained slightly dry-of-optimum, and this is no doubt due to the weakening effect of excess water on the cement hydration. (Similarly, higher strength for dry-of-optimum compaction of sandy soils has been reported by Davidson *et al.*) For lime-stabilised soils (clayey soils) the compaction optima are invariably shifted to higher moisture contents (e.g. Ingles (1962)) and lower maximum dry densities. This lower compacted dry density is *not a disadvantage* of lime stabilisation of clays for two principal reasons (a) it is offset by a greatly increased cohesive strength (and shear strength) in the soil, even when soaked and (b) it possesses the very valuable practical construction advantage of allowing much wider tolerances on construction moisture control, since the compaction curve is appreciably flatter *(Figure 1.3)*. As a general guide, Australian soil testing suggests that specifications for lime-stabilised soils could allow reductions of 0.016 g cm^{-3} per 1% lime added, without loss of performance in pavement sub-bases.

When the criterion of the finished work is soil strength, the relationships between the compaction curves and strength or bearing capacity must be considered. Generally, as was shown by Burmister (1964), *(Figure 1.4)*, the CBR increases markedly with dry density, and this has subsequently been shown also for shear strengths, penetration tests etc. However, it is not a completely general truth. Mitchell (1964) pointed out the *loss* of strength which can occur at higher densities due to changes in clay structure, and Ingles (1971) demonstrated from field trials how an apparently uniform borrow material could give lowest strengths at highest placement densities. The effects involved are as shown in *Figure 1.5*, where comments are superposed on the actual data isobars for a clayey silt.

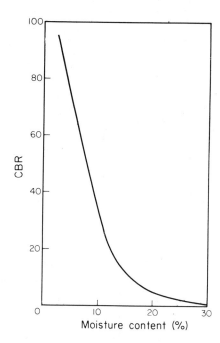

Figure 1.4   Effect of moisture content on CBR

The effects of particle crushing due to rolling may be included as part of a more general *third principle of compaction; that highest densities are achieved by mixtures of different particle sizes.* This principle was established by Fuller and Thompson (1907) who demonstrated the existence of a series of preferred gradings if the highest densities were to be obtained in mechanical mixtures of pavement gravels. The 'Fuller' curves as they are now known *(Figure 1.6)* express the size gradings for materials with various given top sizes which will pack, under a given compactive effort, to the highest densities. The Fuller curves refer to natural materials and mixtures of natural materials, or to unsorted crushed material. Even higher densities can be achieved however, by mixing specially selected, closely monosized fractions of well-rounded particles (McGeary (1961)); even up to as much as 97% of the theoretical density of the solid material. All compaction which relies for its density on good grading of the compacted material is essentially achieved by filling the voids between the larger particles with particles of progressively

6

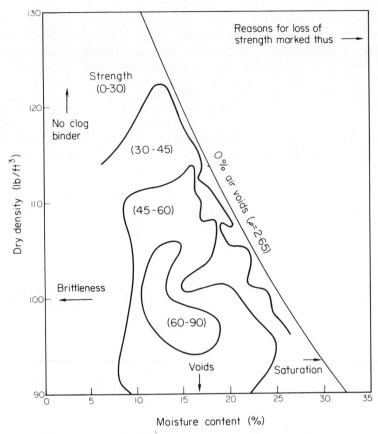

Figure 1.5   Results of field trial showing strength-density relationships

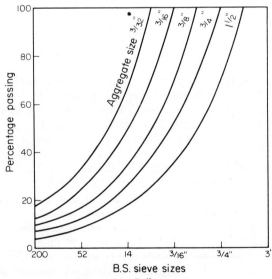

Figure 1.6   Fuller curves

smaller size until even the finest voids have been filled. Note, however, that overworking and reuse of soil can create excessive fines, drop the shear strength and occasionally destroy natural cohesion in a soil such that loss of density occurs by 'over-rolling'.

Bearing in mind this third principle of compaction, it is possible to explain further some of the variations observed in the compaction curves *(Figure 1.2)* of various soils dry-of-optimum. Sometimes the left-hand branch is steep, sometimes flat. Often, at very low moisture contents, a second rise is found. The reasons for such behaviour are now clear. Steep curves arise from strong particles of more or less uniform size. Flat curves arise from particles which crush to a better size distribution than that of the original soil, and for very dry conditions, some clay-bonded soils become extremely brittle by shrinkage cracking and hence may generate fines which, for the given compactive effort, even increase the dry density as the moisture content is decreased.

There is one further essential requirement if any compaction process is to be effective, that is, the existence of a constraining layer around and below the soil being compacted. In practice, this is so important that it should perhaps be stated as a *fourth principle; that the shear stresses in the soil must be confined if compaction is to be achieved.* The confining mass may well be the soil itself, provided the applied compressive forces dissipate within an acceptable distance from the point of application; but where a layer of saturated soil or other soil of low bearing capacity underlies the layer to be compacted, it is obvious that no real compaction can be achieved, because the soft layer will offer no reaction to the load and will deform so as to negate much of the applied compactive effort. In such circumstances, and indeed in all good compaction practice, 'stage' compaction is adopted; i.e. the soil is first rolled with light rollers to develop its own strength, then heavier rollers are applied to complete the compaction. Ideally, this change in bearing pressure can be achieved best by adjusting the ballasting of a roller; or for tyred rollers, adjusting the inflation pressure, ballasting, or both.

Although the four principles discussed above represent the major considerations for soil compaction, many other lesser factors have also been established. These include particle shape, chemical status of the soil water, temperature, and mode of application of the compaction force (such as roller speed, vibration rate), *inter alia.* Their role may be briefly described as follows.

Natural particles vary both in form and in surface roughness. Those particles with a flat and flaky habit normally pack much more densely than those with rounder form (provided the particle size is greater than about 20 microns!). The flat sheets of clay particles and micas possess considerable surface repulsion forces because of their small size, and thus resist compaction so effectively that they fall amongst the most difficult earthen materials to compact. In addition, particles with smoother surface will pack more easily than rough and angular particles due to the lower interparticle friction. (But they are also displaced more easily under shear forces.)

The compaction aid used, be it water or other fluid additive, e.g. bitumen, salt solutions, etc., also affects the response to a given compactive effort in varying degrees and ways. One such important effect is that of various ionic additives in the soil water, the results of which are generally as shown in *Figure 1.7* for clayey soils, due to reaction of the ions with the primary clay particles to cause greater or lesser shear strength in the soil. Where water-immiscible additives are used (such as bitumen) the effect on compaction is broadly the same as for water itself, but for a detailed discussion see Ingles and Metcalf (1972). Increased temperature yields slightly better compacted densities for the same effort (Hogentogler (1936)).

The mode of application of the compaction force has influence in two main ways. Firstly, if it is unidirectional, it will usually produce a compacted state with preferred orientation of the voids i.e. an earthwork whose properties are anisotropic (especially strength, permeability etc.). Secondly, a static force is usually less effective than a dynamic force of the same magnitude since the internal shearing caused by pseudo elastic recovery of

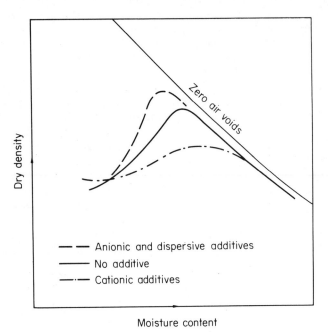

Figure 1.7   *Effect of additives on compaction characteristics*

soil materials under repeated load causes densification in a much shorter time than would be achieved by steady creep under a fixed load. Shear forces provide more rapid compaction than compressive forces, and vibratory forces often provide the most rapid compaction of all.

Indeed, these introductory remarks have not included the influence of time in compaction processes, as it is only in some instances (exemplified presently) within the control of the engineer. Nevertheless it is worth remarking that geological time, by processes of creep, cementation, diagenesis etc. has produced substantial natural compaction in soils and the occurrence of this should be recognised in any field work entailing soil use for construction purposes.

Standard laboratory procedures have arisen for determining the compaction characteristics of soil samples (AASHO and Proctor) and are often referred to as 'modified' and 'standard' compaction respectively. It is common for field construction specifications to call for the attainment of some fixed percentage of these laboratory test values, often either 95% or 100% thereof. As these procedures require large amounts of soil,† a number of miniature tests have also arisen, the most important being the Dietert and Harvard, the results of which do not necessarily match those of the modified or standard procedures, and which must be related to them by experience and practice. One such relationship is given by Ackroyd (*Figure 1.8*) but the variability of soil makes it essential always to verify such relations in the particular circumstances under study. In this respect, the Harvard miniature compactor has considerable advantages, since the pressure exerted by the compaction head and the number of blows and of layers can each be readily varied at

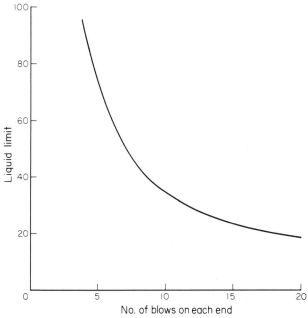

Figure 1.8   Miniature compaction test (after Ackroyd (1957). Courtesy Foulis & Co., London)

will; thus it is necessary to find only a series of parameters which reproduce one given Proctor curve determination, and subsequent testing may be carried out at the miniature level using these matched parameters.

The Proctor test itself consists of dropping a 2.5 kg hammer with 5.1 cm dia. face from a height of 0.30 m onto the soil specimen confined in a 10.2 cm dia. mould. The soil must be compacted in 3 equal layers, each to receive 25 blows of the hammer. The modified (AASHO) test uses a 4.54 kg hammer, and a drop height of 0.46 m; the same number of blows is applied, but 5 layers are specified. Thus the modified test gives a higher compaction than the standard test.

A series of curves (*Figure 1.9*) relating optimum moisture content and dry

† 940 cm³ cubic foot per sample point. To define the full curve at least 5 points are desirable. Moreover, because the soil structure is changed during the compaction process, soil from each determination should never by reused for the next point.

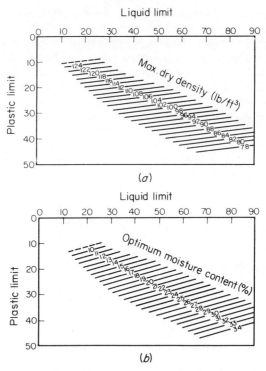

Figure 1.9   Compaction characteristics and index properties

density to index limits for Proctor (standard) Compaction have been published, based on data for a large number of US soils. These charts have also been found useful in Australia as a means of arriving at a first estimate of compaction optima from the simple liquid and plastic limit tests, and are therefore recommended as a cross-check on laboratory figures.

1.2   COMPACTION WITH HEAVY MACHINERY

A recent account of the types of compaction machinery suitable for soil construction work has been given, with illustration, by Ingles and Metcalf (1972), and a detailed review of performance by Morris and Cochrane (1964).

In its earliest mechanical form, compaction was achieved with heavy smooth-wheel rollers ('steam rollers') and the smooth-wheel roller is still a popular unit today. However, many new types of rollers are now available for special soils or for special purposes. These machines include the rubber-tyred rollers, the sheepsfoot rollers, the grid rollers, the vibratory rollers, etc. Their preferred uses are as tabulated on facing page *(Table 1.1)*, and the performance of such rollers has been described in many technical papers such as, for example, Morris and Cochrane (1964), Morris and Tynan (1968), etc.

The most important features of roller compaction are that:

(1) repeated passes of the same roller rapidly lose compaction effect *at the same moisture content* (the increase in densification of a medium heavy

Table 1.1 VARIOUS ROLLER TYPES AND THEIR USES

Roller type	Usual weight (approx.) kg	Usual roll (approx.) dia.	Usual roll width (approx.) m	Contact load kN	Remarks†
Smooth wheel	8100–12 200	1.5 (rear)	1.9	0.35–0.61 per cm width	Better for granular materials. Will operate successfully above O.M.C. (if not too wet).
Grid	6100 unballasted 13 200 ballasted	1.7	1.6	Very high	Should be operated at as fast a speed as possible. Chiefly used for breaking down oversize stones and forcing them below the surface to give a better surface finish.
Vibratory	4050	1.2	1.8	0.21 per cm width	Better for granular materials than smooth wheel rollers; since only half the weight gives equal compaction. Frequency about 2000 Hz cycles/min.
Sheepsfoot	3050–5050 unballasted 5050–7100 ballasted	1.5	1.7	480–1725 per m² (on the feet)	60–120 ft per drum. Best suited for clays, especially in semi-arid zones. The foot pressure may be too high for saturated high moisture clays and too low for very dry clays.‡
Pneumatic tyre	8100–12 200	—	2.3	345–390 per m² (inflation pressure)	4–11 wheels with 9–22 kN per tyre. Best for cohesionless and low cohesion soils, and for surface finishing.

† For roller outputs, see Table 1.2.
‡ A vibratory sheepsfoot is especially good for dry clays.

clay fell off logarithmically with number of passes; Morris and Tynan (1968). This behaviour is not surprising, since increase of density leads to increases of both internal friction and cohesion, hence resistance to further compaction becomes progressively greater for a given compactive effort.

(2) optimum field moisture content for a particular roller may differ from the value determined in the laboratory, and allowance should be made for this fact. In particular the optimum moisture content for the first roller pass may well be slightly higher than for the second pass, and so on similarly; and therefore some drying time can be advantageous between successive roller passes.

(3) the speed of the roller has little effect on the compaction achieved (Markwick (1945)).

(4) a sheepsfoot roller requires more passes than other rollers because of its limited bearing area, and hence poor coverage.

(5) different rollers are better suited to different materials. Thus the kneading sheepsfoot or smooth steel rollers are most suitable for clays, but vibratory, smooth or grid rollers for sands and gravels. In particular vibratory rollers are found to be as effective on sandy soils as non-vibratory rollers of *twice* their weight (Scala (1970)).

(6) the heaviest roller compatible with the restrictions of principle four *ante*, will give the best compaction in every case. (i.e. use of the heaviest roller which does not cause subgrade failures is always recommended.) The change in weight (hence pressure) of a roller can be achieved either by ballasting or by increasing the tyre pressure (for tired rollers).

(7) roller pressure should be greater than the tyre pressures of the earthmoving plant.

Tables 1.2 and 1.3 show the output of various rollers and the compaction standards for heavily trafficked roads (both after Morris and Cochrane). All the above rollers (and the various impact rammers) are designed to compact soil *in layers, at the surface.* Thus they are applied to the construction of roads and embankments, but not to the compaction of deep soils *in situ.* Moreover, they must preferably operate on soils at or below OMC (optimum moisture content) since they are not suited for the compaction of loose or weak saturated soils, in which they are liable to bog. Where they are applied to highway or embankment construction, it has been shown that the effective depth of the compaction produced does not usually exceed 25 cm and hence it is usual to restrict the depth of each new layer to be such that, after passage of the roller, the new layer thickness does not exceed 15 to 25 cm.

Earthen materials for which roller or rammer compaction is unsuited comprise chiefly the weak saturated soils, the highly organic (peaty) soils and some very micaceous soils; in addition to those deep soft soils which must be compacted *in situ.* organic and micaceous soils often possess large elastic recovery (rebound) which makes rolling inefficient. Muscovite, biotite and peat are particularly bad. In most areas, drainage (and at surface level, aeration) will often render soft or saturated soils amenable to roller compaction; but the effect is limited in depth, as is shown by Boussinesq's analysis of the distribution of vertical stress under a surface load, for a homo-

Table 1.2 OUTPUT OF VARIOUS ROLLERS (CUBIC METRE PER HOUR)
(after Morris and Cochrane, 1964)*

Rollers and speed (m/s)	Mod. max. dry density %	Cohesive materials (clays)			Non-cohesive materials (granular)	
		Heavy clay	Silty clay	Sandy clay	Well-graded sand	Gravel-sand-clay
8100 kg steel wheeled roller (self-propelled) 0.9 m/s	85	210	460	280		
	90	90	105	Nil (90)		
	95	Nil (90)	Nil (90)		380	250
	100				90	45
3800 kg vibratory roller (drawn) 0.7 m/s	85	130		170		
	90	25		85		
	95	Nil (90)		Nil (90)	230	170
	100				105	105
3800 kg vibrating tandem roller (self-propelled) 0.7 m/s	85	75		100 (est.)		
	90	34		50		
	95	Nil (90)		Nil (90)	115 (est.)	115 (est.)
	100				60	60
4600 kg taper foot 172 kN/m² sheepsfoot roller (drawn) 0.9 m/s	85	75	165	85		150
	90	Nil (90)	75	Nil (90)		Nil (75)
	95		Nil (90)			
	100					
5050 kg club foot 794 kN/m² sheepsfoot roller (drawn) 0.9 m/s	85	170	150	115		92
	90	100	75	Nil (75)		Nil (74)
	95	Nil (95)	Nil (90)			
	100					
12 200 kg pneumatic tyred 25 kN/m² roller (drawn) 0.9 m/s	85	Nil (85)		305		
	90			Nil (85)		250
	95				130	75
	100				Nil (95)	Nil (95)
20 300 kg pneumatic tyred 55 kN/m² roller (drawn) 0.9 m/s	85	140		460 (est.)		
	90	Nil (90)		145		460 (est.)
	95			Nil (95)	380 (est.)	170
	100				Nil (100)	Nil (100)
13 400 kg grid roller (drawn) 0.6 m/s	85	230		230		
	90	100		70		150
	95	Nil (95)		Nil (95)	150	70
	100				30	Nil (75)

* Figures converted from British units and are rounded off. Figures in brackets are maximum densities after 32 passes (or 16 passes in the case of the 3800 kg vibrating tandem roller).

geneous, elastic, isotropic solid. *(Figure 1.10* shows the change with depth of the vertical stress as a percentage of the applied loading on such a medium). Many refinements of the simple Boussinesq analysis are now available, enabling rapid computation (by finite element analysis) of the stresses and displacements in multilayer systems of various resilient module and resilient strain ratios. Nor are such analyses any longer limited to the homogeneous

14

Table 1.3 DESIRABLE COMPACTION STANDARDS FOR
HEAVILY-TRAFFICKED ROADS (after Morris and Cochrane, 1964)

Cohesive material		Cohesionless material	
Depth below surface, cm	Modified max. dry density %	Depth below surface, cm	Modified max. dry density %
12.7	105	7.6	107
22.9	100	15.2	105
40.6	95	38.1	100
53.3	90	61.0	95
68.6	85	91.4	90
91.4†			

A 5 cm wearing course of asphalt is assumed.
† For high embankments the density of lower layers must be high enough to prevent further consolidation due to the weight of the overlying fill.

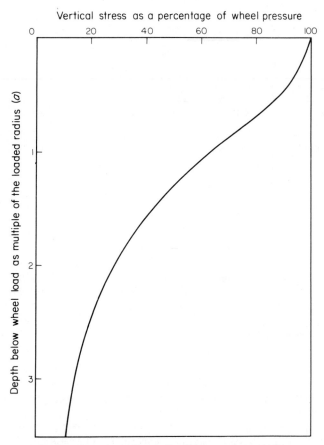

Figure 1.10   Vertical stress distribution on the centre line

isotropic elastic state, but elements of anisotropy and non-linear elastic behaviour may be incorporated. (Miura (1969), Richards (1970).)

If the relevant material characteristics of the soil are known, such analyses may be used to help select the optimum roller for a specific compaction task.

For those soils which cannot be compacted after drainage, or for which conventional drainage is difficult, slow, or impossible; and particularly for soils which must be compacted in considerable depth, other methods than rolling may be applied to effect consolidation. These are of various kinds, and must be discussed separately, as in the following sections.

1.3 COMPACTION BY CHEMICAL MEANS

Reference has already been made to the importance of water as an aid to compaction. Many other compaction aids are also known. In addition to the anionic additives (whose effect is shown in *Figure 1.7*) any addition to the soil which reacts with the clay fraction in such a way as to deflocculate or coagulate (but *not* to flocculate) the clay fraction of the soil, will aid the compaction of a soil dry-of-optimum.

All inorganic salts which act as dispersants for clays (sodium chloride, sodium phosphates, sodium carbonate etc.) are thus a powerful aid to compaction if added to the soil water. The desirable rate of application varies, but generally for the inorganic salts will be between 1% and 2% on the dry weight of soil; whereas for organic additives (e.g. the sulphonic acid derivatives) the rate of addition required may vary from 0.1% to 0.5%. Whilst this group of inorganic and organic (anionic) compaction aids are valuable in dry *clay* soils, a different type of additive may be valuable in *sandy* soils. In this latter case, the soil particles carry little or no surface charge, and dispersing agents are not effective unless some clay cohesion remains. Agents capable of either reducing interparticle friction, or the residual cohesion forces of moisture are, however, effective aids for sand compaction. Such materials are therefore often based either on wetting agents, to ensure good fluids distribution; or on the reverse effect, i.e. emulsions etc. designed to reduce the residual cohesion in the sand by rendering the particles less hydrophilic.

Such methods are generally effective only where mechanised mixing of soil and additive is possible, since only in very permeable soils (sands) will sprays penetrate to any appreciable depth. Other chemical methods are available, however, for soils which cannot be mixed prior to compaction and these methods are based on the unique properties of the quicklime hydration reaction.

This reaction is, as shown, highly exothermic:

$$CaO + H_2O \rightarrow Ca(OH)_2 + 15.2 \text{ kcals}$$

It has other more important advantages though; insofar as firstly, the solid reactants are of very low solubility ($<0.2\%$) and secondly, a substantial density difference exists between the quicklime (3.3 gm cm^{-3}) and the hydrate of lime (2.2 gm cm^{-3}).

It follows that if quicklime is allowed to react with water in a confined environment, substantial pressure (and temperature) are simultaneously generated. This fact can be turned to advantage for the compaction of deep soft soils by the simple technique of drilling or driving holes with hollow casing, filling the holes with quicklime (or if desired for economy, lime–sand mixes may be used), withdrawing the casing and sealing the hole. The heat of the reaction increases the fluidity of the soil water, leading to more rapid extraction from the soil, and the extraction of the water is accompanied by a simultaneous lateral consolidation of the soil under the expansive forces generated by the lime hydrate formed. This method of deep compaction is now practised for major works in Japan and Taiwan, the soils particularly suitable being silts and light loams rather than sands or heavy clays. A full account of the design of such installations is given by Itoh (1969).

Elevation

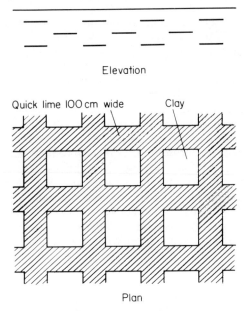

Plan

Figure 1.11   Lime layers

Another method, also based on quicklime, by which accelerated consolidation of a saturated clay may be achieved, is the sandwich method first demonstrated by Yamanouchi (1967). In this method alternate layers of soft clay (up to 70 cm deep) are laid, separated by thin layers (5 cm) of quicklime spread between two thin sheets of special drain cardboard. For better economy, the lime may be laid on a grid pattern in the bank rather than in one continuous layer, see *Figure 1.11*. The drainage of water (which here flows out from the cardboard drains) is speeded both by the temperature and the pressure generated in the reaction, and consolidation is considerably accelerated.

For saturated clay soils *in situ*, the usual method of consolidation is the application of a surcharge weight, usually in the form of a suitable borrow soil. The effect of this surcharge can be increased by providing additional drainage ways by means of spaced vertical holes, filled with sand (for permeability) and connected either to a sand blanket beneath the surcharge soil or else (if available) to a deeper permeable stratum underlying the zone of desired consolidation. A recent variant of these sand drain methods for accelerated consolidation has been the introduction of paper drains, which are 'seeded' into the soil by punching to the required depth and which act as capillary wicks allowing free evaporation at the surface for the deep soil moisture.

Another method for accelerated consolidation of clays is by the application of an electric potential, under which the negatively charged clay particles migrate towards the anode, thus densifying in its vicinity (water can be drained off at the cathode). Although demonstrated in the laboratory (cf. Evans and Lewis (1965)) no extension to field use is known.

1.4   COMPACTION BY SHOCK

Numerous methods for deep soil compaction, based on the application of impact or shock waves have been successfully applied in practice.

The simple extension of the ramming principle to depth may be accomplished either by what has been termed 'direct power compaction', or 'vibroflotation'. The former consists of a driven hollow pile, the end of which is opened or closed, as desired, by remote control so that strata at any selected depth can be compacted by alternately slightly raising and then ramming with the closed head. The mode of operation is shown in *Figure 1.12.*

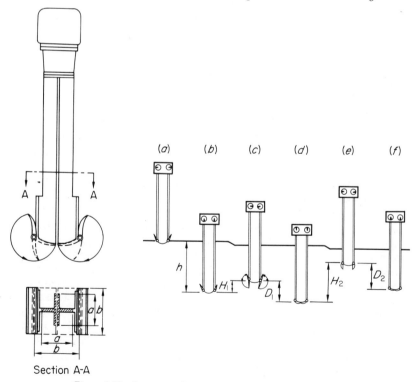

Figure 1.12   Sequence of operations 'Direct power compaction'

Vibroflotation likewise consists of a driven hollow pile, though in this case the initial driving can be assisted by water jets. Having reached the requisite depth the pile is withdrawn slowly whilst gravel or crushed stone is supplied down the shaft and a heavy vibrating action applied to the gravel as it is placed. In both cases, either vibroflotation or direct power compaction, the driving of the hollow shaft itself causes a compaction of the soil, and this effect is increased and intensified through the subsequent impact ramming by the head.

Another such technique is sandpiling, which is rather similar to vibro-flotation but, for economy utilises sand rather than gravel, and relies in part on the drainage characteristics of the sand piles to produce settlement after placement of the columns. *Figure 1.13* shows the type of pile head used. It is best suited for clayey sands and sandy clays where vibroflotation and

Table 1.4   VARIOUS METHODS FOR DEEP SUBSURFACE
COMPACTION AND THEIR APPLICATION

Method	Suitable soils	Borrow materials required	Equipment required
Direct power	Sands	Nil	Special pile driving (a)
Sandpiling	Silts, clays	Sand	Special pile driving (b)
Limepiling	Silts, loams	Quicklime	Special pile driving (b)
Vibroflotation	Sands, silts	Gravel	Special pile driving (c)
Electric shock	Sands, silts	Nil	Special probe driving (d)
Blasting	Sands	Nil	Explosives and probe
Paper drains	Clays, loams	Drain paper and perhaps soil surcharge	Special probe driving (e)

explosives are not appropriate. The three methods are summarised in *Table 1.4*, the direct power method using only *in situ* soil, the sand pile method using sand borrow and the vibroflotation method using gravel or crushed rock borrow.

An alternative method of impact compaction for loose, wet sandy soils is by electric shock. This method, originally developed in Russia, and described in detail by Lomise *et al.* (1963), consists of an impulse generator, high-voltage condenser and electric discharge tip which is inserted in the soil with the aid of a water jet and from which discharges are made at 5–20 second intervals. Each discharge lasts only about one-millionth of a second and in saturated sandy soils very high pressures can be generated (up to $10\,000\,\text{kg cm}^{-2}$) by the shock wave. The impulse generator has a voltage of 60–70 kV and the condenser a capacity of 8 mf. Field work has shown increases of density from, for instance $1.58\,\text{g cm}^{-3}$ to $1.68\,\text{g cm}^{-3}$, and more. This corresponded to increases in relative density from around 30% to about 70%. It is known that soils above about 70% relative density are not subject to liquefaction (cf. Ingles and Grant (1972)) hence all the shock treatments, which essentially rely on liquefaction phenomena, will not produce higher relative densities than this. The radius of effect varies between 1 and 5 m, and costs of around 5 cents per cubic metre are claimed for the process.

Direct application of explosive shock waves has been practised on many soils to achieve compaction, with varying success. The consolidation of sandy soils, especially in dams, has been attempted in this way, as also has the settlement of highway formations across peat bogs. Actually, this latter process, known as 'bog blasting' is not intended to compact the very elastic organic soil, but rather to displace it from under the surcharged fill, allowing the fill to settle into the area from which the organic soil has been displaced. Where sand compaction is attempted by the detonation of explosives, the

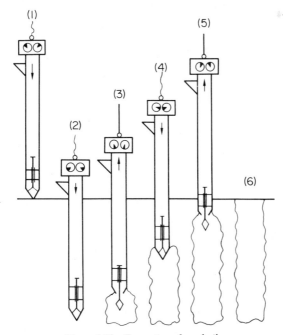

Figure 1.13   Formation of sand piles

presence of silt or clay substantially diminishes the effect. The methods adopted in previous practice have been to explode charges of about 3.6 kg weight at spacings both laterally and in depth of the order of 6.1 m or a little less.

1.5   HYDROCOMPACTION

A form of settlement which has proved extremely troublesome in the field is associated with semi-arid or tropical areas, and occasionally with earlier glacial soils. This arises from sudden collapse, and occurs principally in sandy or coarse residual soils. A short account of its origins is given by Ingles and Grant (1972) and, in general, it may be said to arise from either (a) soils which have weathered to a mechanically metastable condition (e.g.

tropical granitic soils) or (b) soils which have accumulated in a loose condition, for example by wind deposition, in arid areas or (c) rarely, by the action of frost heave with subsequent sublimation of the ice (cf. Prokopovich (1969)).

Soils of types (b) and (c) can easily self-compact (or settle) following irrigation and inundation; although the compaction will of course be aided by surcharge and any other usual means of compaction. This natural collapse following the addition of water to a soil has been termed 'hydro-compaction', and in a sense the 'quick clays' of Norway and Canada which become metastable by leaching also belong to this group of materials liable to self-compaction.

Soils of type (a), being of much more angular grain composition, do not settle so readily on the advent of water, since some degree of mechanical interlock exists. However, as this interlock is progressively weakened by weathering, a situation is finally reached in which a surcharged load which was formerly stable will, when the soil is wetted, cause collapse. Such soils are the Masa soils of Japan, the granitic soils of North Queensland and Rhodesia, etc.

Compaction of all soils susceptible to collapse on water entry is no different from ordinary compaction procedures with the exception that it is absolutely essential that deep profiles of such soils (the commoner case, unfortunately) *must* be compacted in depth, otherwise unpredictable settlements are liable to occur at some future time. In these circumstances, it is essential to recognise soils liable to collapse and to provide against it by either direct hydrocompaction i.e. irrigation or flooding, or any of the methods of deep compaction which can be economically applied in the particular circumstances; preferably aided by vibration to disturb the metastable state. All soils whose natural density is below 1.35 g cm^{-3} should be treated as potentially liable to collapse, though organic (peaty) and volcanic (pumice) type soils which are often of a very low density do not normally suffer hydrocompaction collapse because their friction or cohesion is too high.

1.6   COMPACTION CONTROL IN THE FIELD

Perhaps one of the most important questions to be answered if the engineer is to obtain a satisfactory and predictable performance from his work is, how should compaction be controlled during placement and rolling of the soil?

It should be axiomatic that the most direct measure of the most critical soil property be the means for control, if at all practicable. For instance, in earth dam construction, permeability may be more critical than strength; whereas in road pavements deformation is the most critical quantity. But it is not always possible to measure the most critical parameters by simple, running field tests; so relationships must be established with more easily measured parameters, and the latter used for field control of compaction work. Although this introduces more sources of potential error and greater error variances, the basic design philosophy is that large numbers of tests of a relatively lower accuracy are preferable to a few tests of high accuracy.

This philosophy is probably acceptable for most soils, because the intrinsic variability of soil materials is commonly as much as 20 to 25% (Lumb (1968), chapter 3)). Thus, density has commonly been accepted as a specified control on compaction (usually stated as percentage of modified or standard compaction to be achieved, often 95% or more of the former) though it can be clearly demonstrated (see earlier) that this is far from being a universally reliable measure of performance in service.

An alternative procedure is to prescribe moisture control, and this has distinct advantage over density control in that each moisture content corresponds to a unique density, whereas each density point may represent either of two moisture percentages. However, the rapid field measurement of moisture content is less easily performed than that for density. Occasionally other controls, such as bearing capacity, proof rolling, etc. are prescribed. Some of the more usual tests are listed below.

Control by density
    Sand replacement   (BS 1924:1967 Tests 7, 8.) (American Society for Testing and Materials, 1964, D1556–64)
    Core cutter   Australian Standard AS, A89–1966
    Rubber ballon   American Society for Testing and Materials, 1964, D2167–63T.
Control by moisture content   Speedy Moisture Meter
                                  Neutron Moisture Meter
Other controls
    CBR test   Australian Standard A89–1966, Test 14
    Deflection   Benkelmann Beam
For a complete summary of standard and tentative tests, see Ingles and Metcalf (1972) pp. 342–348.

In the writer's opinion density and moisture content jointly are by far the best indirect compaction control parameters, since together they allow the supervising engineer to determine what residual air voids are still present; and reducing air voids below 5% should be the objective of all good compaction work. Moreover, control on *both* parameters using recovered cores, allows a statistical control of the work more powerful than control on one parameter only, as was shown by Ingles (1971). This is because in the normal laboratory calculation procedure, the errors of measurement in one parameter are compensated by the errors in the other, and a much more accurate control can be attained. (Optimum moisture content and apparent dry density are related by the equation $(100/\gamma_s) - w = K$ (constant) where γ_s is the dry density and w the moisture content. The coefficient of variation of K is found to be very low.)

In all cases, the best means of ensuring good compaction in the sense of reliable and predictable service performance is by statistical control of the testing and acceptance. Metcalf (1969), (1972), Kühn (1971), (1972), Lumb (1971) and others have made important contributions in this field. In particular, it has been shown that outright acceptance and rejection testing is an uneconomic procedure, both for the client and for the contractor. The preferable system is one which provides penalties for sub-standard work rather than outright rejection, the scale of the penalties being prescribed in advance in accordance with known factors such as maintenance costs, loss of service, etc.

A particularly clear statement of the suitability of various specification procedures has been given by Metcalf (1969) and the penalties associated with compliance (as overdesign costs) or non-compliance (as maintenance costs) become quite obvious when it is realised that not all the placed material can be tested, and therefore each test runs an equal risk of condemning good work as it does of approving bad work. (That is, the chance of selecting a bad sample from good material equals the chance of selecting a good sample from bad material.) These risks should be balanced, in the interests of both consumer and producer, whereas current practice frequently ignores consumer risk evaluation and, by theoretically placing all risk in the producer area, ensures overdesign and high costs.

The classical statistical approach to quality control is first to define the parameter variability in terms of a mean value and a standard deviation from the mean. Specifications which demand 'no test inferior to a given value' are particularly inefficient however, for the reasons set out in the preceding paragraph. A better specification is to provide for a pre-estimated proportion defective to a given value, the proportion being determined according to the maintenance and other relevant factors. Ingles and Metcalf (1972) suggest that the following decision rules should be applied in general use:

(1) The material to be rejected outright if more than one out of two samples fails to meet the specification requirements.
(2) The sample should be accepted subject to an appropriate penalty if not more than one out of two fail until twenty samples have been taken.
(3) The material should be accepted outright if not more than four out of twenty fail to meet the specification.

Currently recommended standards for process acceptance and control in practice are as follows:

(1) *Sampling.* Either systematic, random, or stratified random may be adopted, but since the former is open to abuse, the latter two are preferred. Tables of random numbers applied to an imaginary grid overlay will define the sampling points.
(2) *Frequency of test.* As given by *Table 1.5.*

Kühn's method of acceptance is particularly valuable since it relates the

Table 1.5

Product	Property	Process control m^2 per test	Tests per day (min)	Acceptance control m^2 per test	No. of tests (min)
Fill	Density	400–1200	3	175–500	6
	Strength	500–1500	3		
Sub-bases	Density	700–2100	3	350–1000	6
	Strength	1000–3000	3		
	Thickness	400–1200	4	175–500	12
Bases	Density	600–1800	4	250–750	8
	Strength	800–2400	4		
	Thickness	300–900	4	125–375	16

degree of compliance (with specification) to the gain achieved, defined as the difference of product value and product cost. The aim of good quality control work should be to optimise this gain, and Kühn shows how this can be achieved with a continuously graded scale of penalties. Thus, for example, complete acceptance of the work with payment at full tender price is payable when $\bar{x}_n > x_a$. ($\bar{x}_n$ is the mean value of the measured parameter, x_a the acceptance limit specified.) When $x_r > \bar{x}_n > x_a$ the work is conditionally accepted at a reduced rate of payment, P, given by

$$P = \left(\frac{\bar{x}_n - x_r}{x_a - x_r} \right) T$$

where T is the full tender price, and x_a, x_r are the acceptance and rejection limits as predetermined by the designer.

When $\bar{x}_n > x_r$ the product is rejected at no payment. The acceptance and rejection limits can be quantified as follows

$$x_a = \bar{x}\left(1 - \frac{t_{d_a} V_s}{\sqrt{n}} \right) \qquad x_r = \bar{x}\left(1 - \frac{t_{d_r} V_s}{\sqrt{n}} \right)$$

where $\bar{x}$, V_s, are the true mean and coefficient of variation of the product, n the number of observations and t the standard normal deviate (if the distribution of the variable is non-normal more elaborate analysis is necessary, but this is not usually required for compaction work). The quantities d_a and d_r are specified percentages below which the averages of n observations should not fall for acceptance and rejection respectively. d_r will normally be fixed at a low level, say 0.1%, to keep the risk low that an acceptable product will be wrongly rejected. Values of d_a and n must be established either by engineering judgement or economic considerations (both of which *can* be quantified by application of the Bayesian decision theory).

The foregoing remarks describe what seem to be the most valuable methods available for immediate use in compaction control. However, further statistical techniques are already in process of development which hold promise of solution to two most important practical on-site problems: namely, that proper use should be made of accumulated engineering judgement (and conversely, poor judgement penalised) and also that, until such time as very rapid and accurate field tests are available, the speed of construction is such that control testing *must* include a substantial element of undetermined risk. Risk and judgement can now be incorporated into an economic specification of testing and compliance with the aid of Bayesian decision theory, and the definition of such procedures is under study in several centres at the present time.

BIBLIOGRAPHY

ACKROYD, T. N. W. (1957). *Laboratory Testing in Soil Engineering*, Foulis and Co., London.

BURMISTER, D. M. (1964). 'Environmental factors in soil compaction', *American Society for Testing and Materials*, Special Technical Publication 377, 47–66.

DAVIDSON, D. T., PITRE, G. L., MATEOS, M., and KALANKAMARY, P. G. (1962). 'Moisture-density, moisture-strength and compaction characteristics of cement-treated soil mixtures', *41st Annual Meeting, Highway Research Board*.

EVANS, H. E., and LEWIS, R. W. (1965). 'A theoretical treatment of the electro-osmotic consolidation of soils', *Civil Engineering*, October, 1–4.

FUJII, T. (1972). 'Foundation improvement methods for loess soils etc.', *Shikō gijitsu*, 65–74 (in Japanese).

FULLER, W. B., and THOMPSON, S. E. (1907). 'The laws of proportioning concrete', *Trans. American Society of Civil Engineers*, 59, 67, 143, 144–172.

GERRARD, C. M. (1969). 'Theoretical and experimental investigations of model pavement structures', *Ph.D. Thesis*, University of Melbourne.

GREENWOOD, D. A. (1965). 'Vibroflotation', *The Consulting Engineer*, October.

HOGENTOGLER, C. A. (1936). 'Stabilized soil roads', *Public Roads*, 17, 49, 57.

INGLES O. G. (1964). 'The effect of lime treatment on permeability—density—moisture relationships for 3 montmorillonite soils', *Colloqium on Failure of Small Earth Dams*, CSIRO Division of Soil Mechanics, Melbourne.

INGLES, O. G. (1971). 'Statistical control in pavement design', *1st International Conference Applications of Statistics and Probability in Civil Engineering* (Ed. P. Lumb), 267–278.

INGLES, O. G. and GRANT, K. (1972). 'The effects of compaction on various properties of coarse-grained sediments', in *Compaction of Coarse-Grained Sediments* (Ed. Chilingar and Wolf), Elsevier, Amsterdam.

INGLES, O. G., and METCALF, J. B. (1972). *Soil Stabilization*, Butterworths, Sydney.

ITOH, N. (1969). *Seisekkai ni yoru jiban kairyo (Foundation improvement by quicklime)*, Nikan Kogyo Shimbunsha, Tokyo (in Japanese).

KÜHN, S. H. (1971). 'Quality control in highway construction', *1st International Conference Applications of Statistics and Probability in Civil Engineering*, 287–312.

KÜHN, S. H. (1972). 'Quality control in Highway Construction', *National Institute for Road Research (South Africa)*, Bulletin 11.

KÜHN, S. H., BURTON, R. W., and SLAVIK, M. M. (1972). *Economic considerations of quality control for road construction*.

LOMISE, G. M., MOSHCHERYAKOV, A. N., GILMAN, Y. D., and FEDOROV, B. S. (1963). 'Compaction of sandy soils by electric discharger', *Gidrotekhnicheskoe Stroitelstvo*, 7, 9–13 (in Russian).

LUMB, P. (1971). 'Precision and accuracy of soil tests', *1st International Conference Applications of Statistics and Probability in Civil Engineering*, 329–346.

MARKWICK, A. H. D. (1945). 'The basic principles of soil compaction and their application', *I.C.E.*, 3, Road Engineering Division Paper No. 16.

MCGEARY, R. K. (1961). 'Mechanical packing of spherical particles', *Jour. American Ceramic Society*, 44, 513.

METCALF, J. B. (1969). 'Methods of specifying and controlling compaction', *Australian Road Research*, 4, 4–15.

MITCHELL, J. K. (1964). 'Discussion', *American Society for Testing and Materials*, Special Technical Publication 377, 115–117.

MUIRA, Y., and MAKIUCHI, K. (1969). 'Stress and strain analysis of a multi-layered elastic system and its application to the investigation of load response characteristics of the Tomei Highway', *Tsuchi to Kisō*, 17–1, 15–22 (in Japanese).

MORIMOTO, T., FUKUZUMI, R., and ITOH, T. (1970). 'The direct compaction method and its application', *2nd South East Asian Regional Conference Soil Engineering, Singapore*, 189–197.

MORRIS, P. O., and COCHRANE, R. H. (1964). 'Review of the performance of compaction plant on pavement and subgrade materials', *Australian Road Research Board*, Bulletin 1 and Supplement.

MORRIS, P. O., and TYNAN, A. E. (1969). 'The evaluation of pneumatic-tyred, smooth steel-wheeled and vibrating rollers in compacting road materials', *Australian Road Research*, 4, 45–73.

MORRIS, P. O., TYNAN, A. E., and COWAN, D. G. (1968). *Strength, density, moisture content and soil suction relationships for a grey-brown soil of heavy texture*, 4th Conference Australian Road Research Board, 1064–1082.

PROCTOR, R. R. (1933). 'Fundamental principles of soil compaction', *Engineering News Record*, 245–248, 286–289, 348–355, 372–376.

PROKOPOVICH, N. (1969). 'Prediction of future subsidence along Delta-Mendota and San Luis canals, Western San Joaquim Valley, California', *Land Subsidence*, ASH Publication 89, 2, 600–610.

RICHARDS, B. G. (1972). Private communication.

SCALA, A. J. (1968). 'A Study of pressures generated by vibrating rollers', *4th Conference Australian Road Research Board*, 1260–1273.

YAMANOUCHI, T., and MIURA, N. (1967). 'Multiple sandwich method of soft clay banking using cardboard wicks and quicklime', *Proc. 3rd Asian Regional Conference Soil Mechanics and Foundation Engineering, Haifa*, 256–260.

Chapter 2

Stabilisation

O. G. Ingles

2.1 THE NEED FOR STABILISED SOIL

Many years of research in soil stabilisation have resulted less in new techniques (the 'magic wand' still eludes us because of the enormous variety of materials encompassed in the word 'soil') than in better application of existing techniques. Thus lime, cement, bitumen and mechanical admixture still remain the dominant methods by which soil is stabilised in current practice. As conventional materials such as good stone, sand, gravel and the like become scarcer, it is clear that an increasing emphasis must be placed on the proper use of abundant and relatively inexpensive materials such as lime and cement for upgrading the engineering properties of weak soils.

In addition to the pressures of total demand on available resources of good construction material, increasingly high standards of performance are required; and again soil stabilisation is the means by which these demands must be met. Properly applied, it can extend our material resources, improve quality and protect the natural environment.

Soil stabilisation should not be thought of solely in terms of strength; indeed to do so has led, and will lead, to serious structural failures. Soil stabilisation is any process by which some undesirable property or properties of a soil are mitigated or overcome. In engineering practice, there are four chief properties which may require improvement, namely strength, permeability, volume stability and durability (not necessarily in that order of importance). Which of these factors is critical in any given situation depends upon the circumstances of the soil, its environment, and the structure or use proposed.

An extensive discussion of the range of application (and types) of stabilisation techniques has been given recently by Ingles and Metcalf (1972); but generally they may be said to be applicable as follows.

Building foundations. To reduce settlement or heave under buildings either by ensuring volume stability, or controlling permeability or increasing strength. Accelerated drainage, piling, grouting and lime treatments are the most commonly applied, the latter being especially effective against volume instability.

Excavation work. To provide better support in pits, trenches, tunnels etc. by strengthening the surrounding soil or varying its permeability, and to stiffen mine backfills. Methods used in soil are principally accelerated drainage, grouting, anchoring, piling etc. and, for mine fills, cement.

Pavement construction to provide more durable and stronger pavements for highways, airfields, railways etc. by strengthening the soil, to increase its bearing capacity, eliminate volume changes and prevent weathering deterioration. Methods used are chiefly mechanical or chemical (bitumen, cement, lime) admixture.

Slope stability to prevent slips on cut slopes, embankments, natural slopes etc. by strengthening the soil, preventing rock weathering, or varying the soil permeability. Methods applied are chiefly drainage, anchoring, piling, surface seals, etc.

Water retention to ensure safe structures against water erosion (either internal or wave action) especially for dams, tanks, levee banks, canals etc. Methods used are normally either lime or dispersant treatment (according to soil type) to control erosion or reduce the permeability of the soil.

Environment conservation to prevent or repair erosion damage (either of natural surfaces, e.g. gullying; or artificial earthworks, e.g. for drainage ways and abutments), to combat dusting and powdering of road surfaces, land subsidence, chemical corrosion by the soil, etc., by increasing the resistance of the soil to natural weathering from wind or water, and (for earthworks) preventing any rapid deterioration in service. Chief methods in use comprise chemical soil treatments (mostly lime), structural solutions such as step dams, gabions and absorption banks, and vegetational solutions such as the establishment and maintenance of grasses.

Miscellaneous. There are, in addition, many small but important uses of soil stabilisation for purposes such as trench reinstatement, posthole anchoring, and building bricks (especially pisé lime-clay and sand-bitumen or cement).

There are many applications, and also many methods, which will produce a satisfactory result. The choice of any particular procedure will be determined largely by the relative cost of materials in the particular locality. On grounds of cost, it will be usual to prefer a method involving the smallest possible amount of additive necessary to achieve the desired result. However, as will be shown later, this approach entails difficult mechanical problems of mixing large amounts of soil with small amounts of stabiliser (see section 2.4), and also technical problems of ensuring a proper uniformity of quality in construction control (see section 2.6). The latter is especially difficult due to the high variability of natural soil itself. For these reasons, treatments at less than the 2% level have not usually proved economically attractive hitherto (with minor exceptions), and have been one of the compelling reasons for the general use of bitumens, cements, lime and mechanical mixes for soil stabilisation.

The situation is now changing slowly and with new design and control techniques the cost of soil stabilisation should become an increasingly attractive alternative to replacement with other materials or expensive structural design solutions (when available).

2.2 DESIGN CRITERIA AND THE CHOICE OF STABILISERS

The first element in design is to recognise those soil properties which will be critical to a satisfactory performance and service life. This can be done with a minimum of testing—often by field observation only—using soil recognition techniques as described by Ingles and Metcalf (1972). These techniques are intended to optimise the use of local observation and to avoid any call for irrelevant and costly testing, by identifying which soil properties are most likely to prove troublesome for the construction task envisaged.

Ideally, these soil recognition techniques would be coupled with a terrain recognition mapping such as has been developed by Grant (1970). A complete discussion of this integrated approach has been published recently (Ingles (1972)), and may be consulted for detail.

Although the above methods were developed for the Australian and New Guinea environment, they should also be applicable in most other areas of the world not affected by freezing conditions. Areas liable to deep frost require special stabilisation procedures to combat the effect of frost-heave, thawing of permafrost, and like problems.

Ingles and Metcalf's soil recognition technique rests essentially solely on an identification of the clay mineralogy, the texture (grading), moisture status, and natural dispersivity of the soil. From a knowledge of the dominant soil components a reasonable first choice of stabiliser for evaluation can be made, as shown in *Table 2.1.*

Table 2.1 STABILISATION RESPONSE OF MAJOR SOIL COMPONENTS

Dominant soil component	Recommended stabilisers	Reasons
Allophanes	Lime†	For pozzolanic strength and densification
Chlorite‡	Cement	Theoretical (reported stabilisation experience is sparse)
Illite‡	Cement Lime	As for kaolin As for kaolin
Kaolin‡	Sand Cement Lime	For mechanical stability For early strength For workability and later strength
Montmorillonite‡	Lime	For workability and early strength
Organic matter	Mechanical	Other methods ineffective
Sands	Clay loam Cement Bitumens	For mechanical stability For density and cohesion For cohesion
Silts	None known§	——

† Lime gypsum mixtures with gypsum contents up to 40% may be especially favourable.
‡ Thermal treatment is also applicable to most clays if not too wet.
§ Sandy silts may be improved by foamed bitumen, clayey silts by lime or cement.

Table 2.1 is, of course, only a first approximation guide; and in each case some qualifications may be required to its generalised appreciation.

The following specific considerations should be borne in mind for each stabiliser system.

2.2.1   Mechanical stabilisation (mixing of sand with clay or clay with sand, etc.)

Particle size grading should conform either to the continuously graded Fuller curves (Ingles and Metcalf (1972)) or to the specially selected gap-graded mixtures which give high density. Some of the penalties for departure

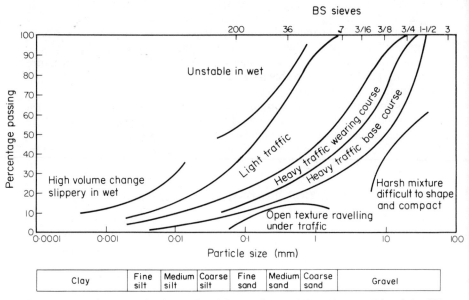

Figure 2.1   Performance related to grading (after Wooltorton (1954). Courtesy Edward Arnold)

from these rules are illustrated by Wooltorton *(Figure 2.1)*. If some cohesion is to be provided for a sand by blending-in clay, then wherever possible the clay or loam chosen should be of a non-dispersive type, since this will often avoid volume change and slipperiness problems.

2.2.2   Cement stabilisation (normal portland cement)

Cement and lime stabilisation, have been discussed in the book 'Soil Mechanics, Selected Topics' (Butterworths, 1968). Large additions of cement are normally considered bad, due to the generation of high strength and hence widely spaced cracking by shrinkage or by service stresses or both (cf. Ingles and Metcalf (1972)). The upper limit for soil work has been considered to be about 8% by weight or 5500 kN/m^2 compressive strength, which ever is the less. Very recently, however, some qualification has been applied to these upper limit restrictions, insofar as it is now known (e.g. Yamanouchi (1965)) that strong soil-cement forms a good layer when placed deep in a pavement structure; and also that for layers nearer the surface there is considerable merit in first forming a strong material then breaking it up by heavy rolling within one or two days after setting. (The actual time delay depends on the strength development and the weight of the rollers available.) At the lower limit, the cement content should be sufficient to reduce soil swelling below 2% and to resist the effects of freeze-thaw or wet-dry cycles without excessive degradation (Ingles and Metcalf (1972)).

Cement should not be used where there is appreciable sulphate or organic matter in the soil, since in the former case subsequent wetting can lead to disruption of the work and in the latter case, strength cannot be developed. In practice, the fineness of cement can also lead easily to segregation in sandy soils, especially when watering is necessary, and mixing procedures should be closely watched for uniformity. A further disadvantage is that compaction of soil-cement work *must* be effected within two hours of mixing if serious loss of strength is to be avoided. (This is due to the rapid cement hydration reaction. Retarded cements exist, but for soil work are generally unsuitable either technically or economically.)

2.2.3 Lime stabilisation (either hydrate, $Ca(OH)_2$, or quicklime, CaO)

Large additions of lime are not desirable in light clay soils, due to loss of cohesion. *For pavement work, about 1% by weight for each 10% of clay content in the soil is recommended, and these amounts should not be exceeded without careful preconsideration (Ingles (1972))*. A small additional amount ($\frac{1}{2}$–1%) should be allowed for inefficient mixing machinery. For anti-erosion work however, the lime percentage should not be reduced below 2% by weight, so as to provide against inadequate distributive mixing. As with cement, lime is not recommended for highly organic soils; but segregation is not a problem since lime is best applied only to clayey soils.

In construction, it must be clearly recognised that lime addition to soil

(1) increases the optimum moisture content for compaction,
(2) reduces the optimum dry density for a given compactive effort,
(3) may or may not reduce the soil plasticity (it can even increase it).

None of these effects is in any way deleterious to the final performance of a lime-treated soil, however. Strength gains more than offset the density losses, and adverse plasticity effects (if any). Indeed, the latter are never good performance criteria for lime-stabilised soil. Positive advantages of lime-soil mixes are the slow but steady gain of strength with time, and especially the easier construction control afforded by the flatter compaction curves.

2.2.4 Bituminous stabilisation (using bitumens, foamed bitumens, tars, cutbacks, emulsion, etc.)

In the bituminous stabilisation of soil, the principal problem is to ensure that excess fluids are not present in the work at the time of compaction, i.e.

the soil to be treated must not be too moist. In this respect, practical difficulties have occurred in areas subject to frequent, high, or irregular rainfall. Moreover, the work must be carefully protected until curing has been effected, and the bitumens chosen to be of a suitable viscosity grade for the environmental conditions of service (hot or cold).

As the important matter of bituminous stabilisation was not discussed in the previous volume (Lee (1968)), it is specifically treated in Section 2.3, and it is sufficient here to note that the principal criteria applied, are those designed to ensure resistance to soaking, provision of cohesion, resistance to swell, and the ability to penetrate and adhere to the soil. There is, for instance, no reason why arid clay soils should not be treated as coarse aggregates suitable for bituminous stabilisation, provided suitable criteria such as these are prescribed and met.

2.2.5   Thermal and other stabilisation methods

Many such systems are known, the most promising being possibly the thermal and the water-setting polymer systems. The advantage of these two, rests in their applicability to a wide range of soil types, but each system also has peculiar disadvantages. For example, thermal treatments are immediately made unfavourable by high moisture content in the soil, because of the high latent heat of evaporation of water. No overall criteria can be set down apart from those which apply to the particular task and the particular stabilisation system.

These special stabilisation methods will normally be chosen only when more conventional stabilisers are inapplicable; as for instance in inaccessible locations (deep stabilisation, tunnelling, grouting, anchoring etc.) or where particular soil requirements must be met, as for instance with peats, silts and other soils troublesome with conventional treatment methods.

2.3   PROPERTIES OF BITUMINOUS MATERIALS AND THEIR APPLICATION IN SOILS

A wide range of bituminous materials have been applied to the stabilisation of soil, and include petroleum bitumen, petroleum tar, coal tar and various preparations based on these such as foamed bitumen, bitumen emulsions, cutbacks, latex impregnated sprays, heavy and light oils, etc.

They do not react with the soil or its components, and their effectiveness as soil stabilisers rests wholly on two properties; their disaffinity for water (i.e. waterproofing ability) and their viscous cohesion. In a cohesionless soil, bitumens provide an alternative source of binding which possesses certain important advantages over the cheaper clay binder of a mechanically stable mix. These are, chiefly, that their cohesive effect is not sensitive to changes of water content after construction, and that they have better internal viscosity and rheological properties for flexible behaviour (clays tend to exhibit either a low shear resistance or a brittle behaviour rather than a wide range of viscous flow behaviour).

It is useful to look in more detail at the two main advantages conferred by bituminous additives.

(i) Waterproofing action. Diffusion of water through bituminous or oiled films is slow, but finite. They are not complete water barriers; a bituminous spray over clay will not prevent, but will only delay, its drying out in a low humidity environment. Addition of soluble salts or lime can increase this permeability; although where a good aggregate is present, the lime will in time form a cemented structure with greatly improved bonding. The permeable nature of a bituminous-bound soil is shown by results such as in *Figure 2.2*. However, it is often not necessary to rely on long term water resistance, since for pavements (for example) it is necessary to rely on water-resistance only over the period of a wet season at most, and often only over the period of a few days at a time. Bituminous stabilised soil is effectively proofed against the entry of *bulk* water, but in general is not impervious to water *vapour* movement and slow diffusion processes.

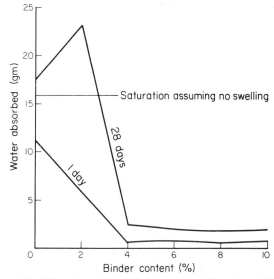

Figure 2.2   Effect of qualtity of binder on permeability of a stabilised soil

There is evidence that the waterproofing properties are improved by higher temperatures at the time of application (Justo and Dayal (1967)) and that they are conferred equally on heavy clays as on sands (Morris (1970)). Indeed, pavement sections on bituminous treated expansive black soils have behaved as well or better than any other comparable test section.

(ii) Cohesion and adhesion. The internal cohesion of bitumens is high at normal temperatures, but embrittlement occurs at lower temperatures and the cohesion falls off at high temperatures. Moreover, exposure leads to oxidation and this in turn leads to embrittlement. A durability test is therefore always relevant to the performance of bituminous binders.

The overall strength of a bituminous bound soil derives chiefly from the matrix rather than the binder, hence an optimum percentage bitumen exists below which strength falls off due to insufficient cohesion in the soil mass, and above which the thicker films of bitumen lead to loss of strength.

   The adhesion of a bituminous film to the soil or aggregate is important, not only for cohesive strength of the whole treated mass, but also for ensuring better water repellency, since natural soil surfaces are normally water-affinitive. Many methods have been adopted to ensure good adhesion, one of the best being the addition of 1–2% lime (as hydrate) to the bitumens. Lime has a strong affinity both for the soil surfaces and the bitumen phase. The mechanism of the latter is not well understood, but is presumed to be a reaction with the bitumen/tar acids. Other methods adopted to improve adhesion include the use of cationic additives; but the quantity must be carefully regulated to achieve the desired effect, and their addition to the soil rather than to the bitumen may be desirable (Ingles and Metcalf (1972)). Stripping, due to insufficient bond within or under the stabilised layer, is a

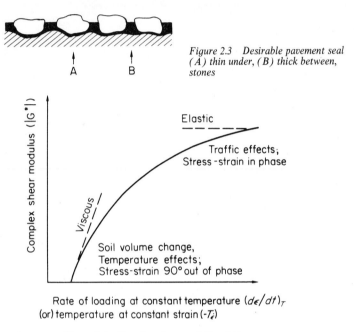

Figure 2.3   Desirable pavement seal (A) thin under, (B) thick between, stones

Figure 2.4   Complex shear modulus of bitumen

familiar problem with bituminous stabilisation, especially in clean sands; and care should always be taken to ensure good adhesion and penetration, if necessary by precoating the soil at higher temperatures.
   Because the requirements for strength (cohesion/adhesion) and water-proofness in a bituminous stabilised soil are to some extent not mutually compatible, great care must be exercised in choosing an additive percentage. This technical problem may be illustrated as in *Figure 2.3* which represents the desirable properties of a flush seal for a pavement, namely a thin film *under* the stone (for strength) and a thick film *between* the stones (for waterproofing and to minimise oxidation).
   A description of the typical material specifications and tests for bituminous stabilisers is given by Ingles and Metcalf. No tests are wholly satisfactory, and even the penetration index test is now becoming obsolete because of the

difficulty in interpreting results obtained half way between purely viscous and purely elastic behaviour. Although the actual mechanisms of the stabilisation action of bitumens in soil are known only in rather broad terms as described above, nevertheless the rheological properties of the bitumens themselves are much better defined. Dickinson (1972) has pointed out that a complex shear modulus for bitumens can be defined, whose magnitude is dependent on the rate of strain and the temperature. (For small strains, the response in the ambient temperature range is linear. The elastic modulus of bitumen is approximately 10^6 kN.m^{-2}).

This complex modulus is defined as the ratio of the stress to the strain and, in general, these are out of phase. *Figure 2.4* illustrates the situation and its practical implications. Dickinson has further suggested that the modulus of a bituminous bound aggregate or composite may be described by the equation

$$E_T = 3|G_s| \frac{\pi}{4} \cdot \frac{r_s}{d} \tag{2.1}$$

where r_s is the mean equivalent sphere radius of the aggregate;
  d is the closest spacing surface to surface of aggregate units;
  G_s is the complex shear modulus of the bitumen binder;
  E_T is the tensile modulus of the stabilised mass,
  both G_s and E_T are measured at the same rate of strain and temperature.

2.3.1 Penetration of bituminous materials

Apart from the direct application of hot bitumens, with or without various types of inert filler, colder treatments can be successfully undertaken using emulsions or cutbacks. The former depends on dispersing the bitumen in water with subsequent time-controlled breaking of the emulsion, and the latter is based on dissolving the bitumen in a volatile oil (kerosene or diesel fuel is usual) with subsequent evaporation of the solvent. Both these methods ensure a better penetration into dense soils than can be achieved with bitumen alone; hence they give some protection against stripping; but both require curing.

A recent development of considerable interest is the use of foamed bitumen, formed by blowing steam through hot bitumen of a grade of R90 or harder. This foamed product is much more effective for coating damp aggregates than either emulsions or cutbacks and is economical of the bitumen. It must be remembered however that, the higher the temperature at which the bitumen is blown, the durability will be reduced accordingly.

Antioxidants and deoxidants for hot bitumens have been proposed and tested (Zinc diethyl dithiocarbonate is one such) but are still essentially in the research stage.

2.4 THE MIXING OF STABILISERS AND SOIL

Although some forms of soil stabilisation can be effected by infiltration, injection, layer spreading and similar means, the mixing procedure is the most common. This is because it is usually desirable to improve the whole volume of the soil uniformly, rather than to have point-to-point variation in the mass properties (for example, with piling techniques) and secondly, because unassisted infiltration into many soils is a slow and inefficient process, if it occurs at all (even at high pressures, many soils cannot be effectively grouted). The simplest way to promote uniform blending of soil and stabiliser is by some form of mechanical mixing. Many commercial machines have been developed to do this (an illustrated description is contained in Ingles and Metcalf (1972)) and the present discussion will be limited to examining the basic requirements which must be met to ensure good work.

One of the earliest studies was that of Smith (1955). A more general investigation has been reported recently by Ingles, Metcalf and Frydman (1966), who defined the concept of an 'adequate mixture'† as follows:

'n substances are defined to be adequately mixed when there is an acceptable probability that the smallest volume of subsequent practical importance everywhere (in the whole mixture) contains each of the n substances within acceptable limits of the predetermined proportions'.

Note that if the 'smallest volume of practical importance' is judged by the engineer to be of cubic metre or larger size, the discrete piling stabilisation systems become acceptable; and where the volume is judged to be sub-microscopic, a liquid infiltration system is necessary. In short, the common usage of mechanical mixing for stabilisation simply reveals a judgement (perhaps often subconscious) on the part of the engineer that the smallest volumes of practical importance lie in the coarse sand size range since this is the usual product specified of modern machinery. Most standard specifications call for '80% passing 4.7 mm sieve' or equivalent requirements. In some degree this probably reflects the long established practice of using coarse aggregates in construction work rather than soils, but in the recent literature there is much accumulated evidence that particles of millimetre size upwards dominate the mechanical behaviour of a soil, and hence justify the adoption of mixing processes for soil stabilisation (Marshall and Quirk (1950); Ingles (1963); Lafeber (1966)).

A mechanical mixing process consists of two (independent) actions namely, distribution of the components and subdivision (often termed pulverisation) of the original materials. Each material to be mixed will have different distributive and subdivisional properties which must be provided for by the mixing method adopted. For example, fine sized material such as cement may be difficult to distribute uniformly in a coarse sand even though both components are readily subdivided, whereas the subdivision of a moist heavy clay is an extremely difficult practical problem even though cement can be distributed uniformly across it.

In such circumstances, it is relevant to consider whether distribution and subdivision are equally important to the final product, or whether one is more important than the other. Ingles, Metcalf and Frydman concluded on

† Because of the impreciseness of the concept 'mixture' as regards randomness or order, and the fact that soils are themselves heterogeneous, a more practical conceptual term directly related to end use, and measurable, must be adopted for soils work.

the basis of laboratory and field tests that, although the answer to this question depends on the nature of the stabiliser itself, for those stabilisers in common use an adequate distribution is much more important than good subdivision. These conclusions are borne out by field experience (as for instance in the bituminous stabilisation of heavy clays (Stocker (1972)). However, the actual cost of power for pulverisation by modern machinery is small, and the desirable mode of operation for practical stabilisation appears to be rapid high power input single-pass mixing with particular attention paid to the uniformity of spread of the stabiliser before blending.

At the present time, the uniformity of spreading is often less than satisfactory. Kühn (1972) reports a coefficient of variation of 14% for field work in South Africa, and values as high as 50% have been reported in Australia (Ingles *et al.* (1966)). In the main, this latter value reflects the

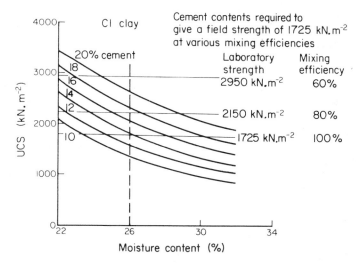

Figure 2.5   Effect of mixing (after Robinson (1952))

problem of blending small amounts of stabiliser with large amounts of soil. The uniformity achieved becomes appreciably poorer the more disparate the quantities of the components to be blended. Since economy of additive is important in soil stabilisation, where the stabiliser itself may represent 30% or more of the total cost of the work, there exists a real need for mixing processes which ensure better uniformity of distribution at low percentage additive levels.

One such method, applicable in dry regions, is the application of stabiliser in a slurry form, e.g. lime slurries; and for wet soils there may be merit in bulk spreaders using air-lift, i.e. fine fluidised solids application.

The foregoing remarks do not mean that mixing cannot be achieved by simple, or inefficient means. These are perfectly allowable *provided the cost penalties for inefficient processes are properly understood.* A good illustration of this has been provided by Robinson *(Figure 2.5).*

Considering the four main soil stabilisation systems in turn, mechanical mixing poses few serious problems since, in general, substantial quantities of additive are involved and the objective of a good grading minimises any likelihood of segregation or pulverisation problems.

Cement blending must be watched for segregation in open sands, which may be accentuated by water additions. Water must be regarded as an equally important mix component, to be uniformly distributed as far as possible. In heavy clays good blending requires good pulverisation of the clay. But it is important to bear in mind, for cement work, that an hydration reaction proceeds rapidly, as soon as the cement enters the soil and therefore the time allowed between mixing and compaction can critically alter the properties of the stabilised mass. Strength, for instance, will be greatly diminished if this time exceeds 2 to 3 hours.

Lime blending (which occurs mostly with clayey soils, since lime is not a stabiliser for clean sands) has fewer problems than cement. But, for heavy clays, it is sometimes found expedient to mix in two stages, the first stage consisting of a low percentage lime treatment which, if allowed to cure for one or two days, makes the subsequent pulverisation of the soil easier and better for mixing-in the final lime (or cement) addition.

Bituminous stabilisation, when it is effected by spraybars and tining, faces only the problem of mixing in depth. Because of its high viscosity this can prove troublesome, and streakiness or apparent segregation are not uncommonly observed in recovered specimens. Provided soil aggregations are coated however, this does not appear to affect the overall performance greatly and because of the different mechanisms involved in bitumen stabilisation as compared to cement, lime, etc. (vide 2.3 *supra*), it seems likely that good mixing, whilst always desirable, is less critical here—or can tolerate a much larger 'smallest critical volume'—than for other forms of stabilisation. This is an area in which further research would seem to be appropriate.

2.5   GROUTING WITH STABILISERS

An alternative to the mixing process for soil stabilisation is the grouting procedure, adopted whenever it is impracticable to disturb the soil, e.g. in tunnelling or open excavation.

Grouting of soils is practised either to reduce water movements, or prevent heave or subsidence, or to stiffen weak or cohesionless strata. Because it is a comparatively expensive operation and hence used only in special circumstances where no good alternatives exist, it has remained until quite recently—and in many respects still is—more of an art than a science. Recent developments in equipment, in grouts, and in the understanding of their function and use make it appropriate to outline the principles governing the stabilisation of soils by grouting.

To be effective, a grout must penetrate the soil thoroughly, set or solidify *in situ*, and be permanent. The ideal requirements for a soil grout might be said to be small particle size (to ensure penetration), low viscosity (also to ensure penetration), permanence, strong adhesion (for strength) controllable set (to ensure correct placement), no syneresis (to ensure a seal) and no toxicity. Also, naturally, low cost! No grout meets all these criteria, and an objective choice has therefore been difficult to achieve.

The first grout ever used appears to have been lime, by Bérigny in France in 1802. Cement grout was first introduced in Germany in 1864. Sixty years later the Dutch engineer Joosten applied the first chemical grout (sodium silicate–calcium chloride). Nowadays a multitude of grouts exist, many of which are patented proprietary mixtures but all may be classed into four groups. This grouping, as given below, tends to be a price and performance grading, in which cost and effectiveness both rise in the direction (1) to (4).

(1) Suspension or particulate grouts, e.g. cement, lime, etc.
(2) Emulsion and colloidal grouts, e.g. bitumens or bentonite.
(3) Two-solution chemical grouts, e.g. sodium silicate–calcium chloride etc.
(4) One-solution chemical grouts, e.g. polyacrylamides, polyurethanes etc.

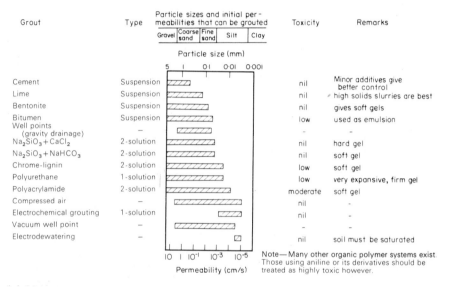

Figure 2.6   Applicability of grouting

The present discussion will be limited to a review of some grout properties and their significance. For a recent treatment of the means of grouting in practice, Ingles and Metcalf (1972) or Ingles (1972) may be consulted.

The requirement of small particle size for a grout applies only where it is desired to inject fine-grained clayey soils. Where large open fissures are to be sealed, then the coarsest size grout available is likely to be both most effective and most economic. *Figure 2.6* shows the penetration sizes for various grouts.

Viscosity is equally as important as particle size, however, for the two reasons that

(1) low viscosity ensures good penetration into soils, especially fine-grained soils and,
(2) high viscosity ensures economical grout usage when it is desired to close off large fissures in a soil or rock.

Most chemical and emulsion grouts are either newtonian or nearly so in their rheological properties; suspension grouts however are non-newtonian, often to a marked degree, and this acts as an additional factor limiting their penetration whenever the shear stress in the grout fluid falls below the threshhold value for flow of the particular grout. An account of this effect has been given by Raffle and Greenwood (1961) for cement and bentonite grouts and by Ingles (1970) for lime grout.

Little data are available to indicate the permanence of the various grouts, but cases of expulsion after placement are known and, if high water pressures are to be resisted, it is presumably wise to choose a grout with good adhesion to the soil. This, in turn, has not yet been measured in any serious quantitative sense, but certain qualitative observations, e.g. Ingles (1972) have been made which indicate a particularly strong adhesion to silica surfaces by the polyurethanes, and this appears to be corroborated in strength data recently presented by Vinson and Mitchell (1972).

Accurate control of the gelation time of a chemical grout often requires very accurate mixing and proportioning equipment (which should also be capable of a high volume ratio of delivery). Suspension grouts rely for their effect on the subsequent separation of solid and liquid phases by a kind of 'filtration' effect, in which the pore channels of the soil become progressively blocked by sedimented grout solids, which proceed to filter out finer and finer solid fractions from the injected suspension. To prevent premature sedimentation, protective colloids such as bentonite (up to 5%, but usually much less) may be added to suspension grouts; and to accelerate sedimentation blockage in very coarse pores 'bulking out' of the suspension with sand, sawdust, or other cheap fillers is often adopted. As an illustration of the filtration effect, it has been shown (Ingles (1972)) that when injecting a colloidal cement suspension in which the largest particle size was 20 microns and the mean size near to 8 microns, only water flow occurred into plates set at a spacing of 10 microns.

For practical economic reasons (to minimise drilling and injection costs), it is usually desirable to seek a maximum radius of penetration from each injection point. Various formula have been proposed to enable the radius of penetration of a grout to be computed. One simple formula, based on spherical distribution of the grout is

$$r = \sqrt[3]{\left(\frac{3qt\beta}{4\pi e}\right)} \tag{2.2}$$

where q is the grout take in unit time, t the time of injection, β an expansion factor ($\beta = 1$ for non-expansive grouts), and e the void ratio; but a better equation is one based on the work of Raffle and Greenwood

$$\frac{t_1}{t_2} = \frac{\frac{n}{3}(p_1^3 - 1) - \frac{n-1}{2}(p_1^2 - 1)}{\frac{n}{3}(p_2^3 - 1) - \frac{n-1}{2}(p_2^2 - 1)} \tag{2.3}$$

where $p_x = r_x/r_o$, and n has values 2–3 for most grouts. r_1, r_2 represent the radii of penetration of the grout at times t_1, t_2 respectively. The formula must not be used, of course, for times in excess of the cut-off time appropriate to the shear strength of any non-newtonian grout. For such grouts, permeation cannot reach further than a radius R_L given by

$$R_L - a = \frac{\gamma_w g h \alpha}{2S} \tag{2.4}$$

where S is the shear strength of the grout, α a constant depending on the soil permeability and a the radius of the injection hole.

For a treatment of the general practice of grouting, Fujii (1969), Ishy and Glossop (1962), or Karol (1960) may be consulted.

2.6   TESTING AND CONTROL OF STABILISED SOIL WORK

As has been indicated in discussion of the mixing of soil with stabilisers, the quality of the work depends on the uniformity of the mixture achieved; and with substantially disparate quantities involved, problems of quality control and testing specification arise.

Certain testing specifications already exist, and have been shown in Chapter 1 *(Table 1.5)*, but these do not cover testing for the proper quantity and distribution of stabiliser in the soil. A number of substantial problems make such testing difficult: firstly, almost all tests for cement, lime, or bitumen content are slow and costly, hence they cannot be used in any meaningful way for the control of work *as construction proceeds*. Secondly, if the work is allowed to cure for any length of time, reactions occur which may make the subsequent chemical analysis difficult or even impossible to interpret, especially since the soil itself often contains components similar to the lime or cement themselves.

In these circumstances it is preferable to base the quality control for construction and acceptance on some essential property of the stabilised soil which can be readily measured in the field; preferably one which has a low coefficient of variation in the sampling and testing of the soil. Ingles (1972) discussed some aspects of this choice as regards pavement work, and indicated that preferred choices might be compaction voids (calculated from density and moisture content) or a simple strength test (because of the very low Poisson ratio of stabilised soils, *circa* 0.1, an unconfined compression test is quite suitable). However, thickness control is also necessary in stabilised soil work which involves mixing and laying procedures, hence a test such as the California Bearing Ratio which monitors the effects of both intrinsic strength and thickness in the one measurement is probably preferable. Little data on its coefficient of variation are available at present, although published figures indicate a value of about 10–25%, which is comparable with that of most other measurable parameters. For CBR testing, the process control and acceptance specifications would be as set out in *Table 1.5* for 'strength'.

For obvious reasons, in any testing programme, it is usually better that the selection of test points be made according to some system of random allocation rather than in any regular pattern.

As Metcalf (1972) has pointed out, the risks of accepting sub-standard work, or of rejecting above-standard work, may be defined according to the equation

$$P = \sum_{0}^{r} \frac{n!\,p^r q^{(n-r)}}{r!(n-r)!} \tag{2.5}$$

where P is the probability of acceptance (i.e. the risk of getting not more than r defectives in a sample of n with a proportion defective p $(q = 1-p)$). The consumer's risk of having poor quality accepted and the producer's risk of having good quality rejected may be set at any desired level by selecting appropriate values of P and p. Obviously, this is a much more desirable procedure than specifications such as 'no test shall fall below the minimum', as can be seen from *Figure 2.7*, although the proportion defective which is economically acceptable must be defined (Kühn's method, see Chapter 1, is suggested for meeting this requirement).

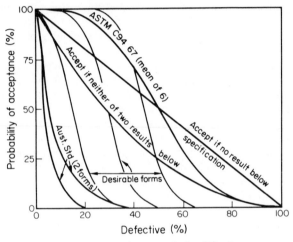

Figure 2.7   Quality control of stabilisation

The total variability observed in a stabilised soil arises from many sources —variability of the soil, variability of the stabiliser in the soil, variability in the sampling, and variability in the testing procedures. None are negligible, and McMahon (1970) found that the variability of compaction density test results was distributed in the proportions 40–47% for materials, 10–27% for sampling, and 25–50% for testing. In these circumstances, the variability of the stabiliser in the soil is not necessarily a major component in the total variability observed from test results, and precautions must be taken to ensure that inadequate work is not accepted without penalty. This may take the form of a testing compliance procedure such as embodied in equation (2.5), or alternatively an initial quantity specification based on known variability parameters for the soil. The latter course may seem difficult, but in practice the sampling and testing variability of a standard or regularly used laboratory will be known with some accuracy, and the variability of the soil itself will be known from the initial planning tests. Moreover, as the other coefficients of variation involved are all substantial, the influence of the variability of the distribution of stabiliser in the soil on the total variance is not great. If desired, a safety factor might be incorporated at this point.

This general method is supported by the relative insensitiveness of the magnitude of the scatter to changes of the additive percentage i.e. the experimental coefficient of variation tends to increase as the additive percentage is decreased. Using the above estimate in conjunction with the desired minimum level of additive and a prior assessment of the total area which can be tolerated defective to that level (a choice similar to that involved in mixing to the 'least volume of subsequent practical importance') an additive percentage can be specified for the work. Given normal supervision, this will not require control testing other than what might be necessary to prevent slack work. This approach can be summarised as follows

$$\mu = L + k\sigma \qquad (2.6)$$

where μ is the mean, σ the standard deviation of the mass of treated soil, and L that value below which a certain proportion p of the mass lies. For a normal distribution, values of k corresponding to chosen values of p may be obtained from the appropriate single tailed distribution *(see Table 2.2)*.

Table 2.2 PERCENTAGES (p) EXCEEDING k TIMES THE STANDARD DEVIATION FOR A GAUSSIAN DISTRIBUTION

p	0.01	0.05	0.1	0.5	1	2.5	5	10
k	3.70	3.29	3.05	2.58	2.33	1.96	1.65	1.28

Using the input values of p, σ, and L, a value for the design mean can be estimated. For example, for a coefficient of variation of 25%, an accepted proportion defective of 1%, and a minimum desired value (L) of 2% stabiliser in the product, $L = 2$, $\sigma = 2 \times \frac{25}{100} = 0.5$, $k = 2.33$ *(from Table 2.2)* and hence the design should specify $2 + 2.33 \times 0.5 = 3.17\%$ stabiliser. The substantial increase over the desired minimum level of 2% simply reflects the high quality specification (less than 1% defective) and variability of the construction techniques adopted. If, for instance, the costs of maintenance were such that a 10% defective level could be tolerated, then the prescribed additive level would become only 2.64%. Even with maintenance costed at penalty rates, the saving in stabiliser quantities may justify a higher level of defectives acceptance.

In fact, a minimum cost curve can be constructed based on given values for maintenance and stabiliser, from which an optimum level of defectives acceptance can be objectively chosen. *Figure 2.8* illustrates the case, using the numerical values from the previous example, and assuming two cases, one in which maintenance costs four times as much as the stabiliser, and one in which it costs only twice the stabiliser. In the latter case a fairly wide choice of percentage defective gives little change in the overall costs (in such a case, presumably the lowest percentage defective will be chosen), but in the former case a fairly sharp optimum saving in the total costs occurs at about $1\frac{1}{2}\%$ defectives.

Whether this type of approach to quality assurance is better than the prescribed testing programmes outlined earlier depends essentially on whether reliable coefficients of variation are available either from previous tests on the same site or from accumulated file data, and emphasises the importance of ensuring maximum efficiency of data storage and recovery procedures.

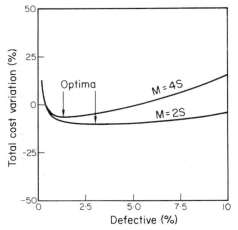

Figure 2.8   Effect of quality control on cost

BIBLIOGRAPHY

DICKINSON, E. J. (1972). Lecture 9 in *Towards New Methods in Highway Construction*, School of Highway Engineering, University of New South Wales.

FUJII, T. (1968). *An Introduction to Rational Grouting*, Tokyo.

GRANT, K. G. (1970). 'Terrain Evaluation for Engineering Purposes', *ARRB Special Report No. 6 Symp. on Terrain Evaluation for H'way Engng.*, Townsville, Sept., 81–105.

INGLES, O. G. (1972). *Tunnel Grouting*, (in preparation).

INGLES, O. G. (1972). Lectures 5 and 6 in *Towards New Methods in Highway Construction*, School of Highway Engineering, University of New South Wales.

INGLES, O. G. (1963). 'The Shatter Test as an Index of Strength for Soil Aggregates', *First Tewkesbury Symposium Proceedings*, Faculty of Engineering, Melbourne University, Butterworths, 284–303.

INGLES, O. G., and METCALF, J. B. (1972). *Soil Stabilization*, Butterworths, Sydney.

INGLES, O. G., METCALF, J. B., and FRYDMAN, S. (1966). 'The Mixing of Stabilizers with Soil', *Civil Eng. Trans. Inst. Engrs. Australia*, 203–213.

INGLES, O. G., and NEIL, R. C. (1970). 'Lime Grout Penetration and Associated Moisture Movements in Soil', *Aust. Inst. Engrs. Symposium on Soils and Earth Structures in Arid Climates*, 28–32.

ISHY, E., and GLOSSOP, R. (1962). 'An Introduction to Alluvial Grouting', *Proc. I.C.E.*, London, **23**, 705–725.

JUSTO, C. E. G., and DAYAL R. (1967). 'Effect of Mixing and Aeration on Soil-Cutback Mix Design', *Indian Roads Congress*, **32**, 444–456.

KAROL, R. H. (1960). 'Chemical Grouting' in *Soils and Soil Engineering*, Prentice Hall, Englewood Cliffs, N.J.

KÜHN, S. H. (1972). 'Quality Control in Highway Construction', *National Institute of Road Research (South Africa)*, Bulletin 11.

LAFEBER, D. (1966). 'Soil Structural Concepts', *Eng. Geol.*, **1**, 261–290.

LEE, I. K. (Ed.) (1968). *Soil Mechanics, Selected Topics*, Butterworths, London.

MARSHALL, T. J., and QUIRK, J. P. (1950). 'Stability of Structural Aggregates of Dry Soil', *Aust. Jour. Agric. Res.*, **1**, 266–275.

MCMAHON, T. F. (1970). 'Quality Assurance for Highway Construction in the United States of America', *Proc. 5th Conf. Australian Road Research Board*, **5,** 195–206.

METCALF, J. B. (1972). 'Quality Control for Construction', Lecture 10 in *Towards New Methods in Highway Construction*, School of Highway Engineering, University of New South Wales.

MORRIS, P. O. (1970). *Progress Report of an Investigation into Road Sections in Queensland*, Aust. Road Res. Bd., Internal report.

RAFFLE, J. F., and GREENWOOD, D. A. (1961). 'The Relation between the Rheological Characteristics of Grouts and their capacity to permeate soil', *5th Int. Conf. SMFE, Paris*, **II,** 789–793.

ROBINSON, P. J. M. (1952). 'British Studies on the Incorporation of Admixtures with Soil', *Proc. Conf. Soil Stabilization*, M.I.T., 175–183.

SMITH, J. C. (1955). 'Mixing Chemicals with Soil', *Industrial and Eng. Chem.*, **47,** 2240–2244.

STOCKER, P. (1972). Private communication.

VINSON, T. S., and MITCHELL, J. K. (1972). 'Polyurethane foamed plastics in Soil Grouting', *Jour. Soil Mech. and Fdn. Eng. ASCE*, SM 6, 579–601.

WOOLTORTON, F. L. D. (1954). *The Scientific Basis of Road Design*, Edward Arnold, London.

YAMANOUCHI, T. (1965). 'Effect of Sandwich Layer System of Pavement for Subgrades of Low Bearing Capacity by means of Soil Cement', *Proc. 6th Int. Conf. SMFE, Montreal*, 218–221.

Chapter 3

Application of Statistics in Soil Mechanics

P. Lumb

3.1 INTRODUCTION

Civil engineering design is largely a matter of decision making under uncertainty. The loads and responses are never known exactly, the accuracy of the design method is uncertain, and the choice between alternative designs or construction procedures is rarely clear cut. At all stages from initial concept to final completion, decisions must be made using either incomplete information or super-abundant and mutually inconsistent information. This inherent uncertainty is particularly characteristic of soil engineering since natural soils are extremely variable in their properties and the rational choice of suitable design parameters is generally the most difficult part of a design for an inexperienced engineer. Little guidance can be found in the standard soil mechanics textbooks or codes of practice.

Statistical methods can be of great value to the designer since it is possible to express many of the decision uncertainties in terms of numerical probabilities, thus allowing quantification of judgement to some extent and clarification of the problems. Decisions cannot be avoided, of course, but by incorporating the variabilities, uncertainties, and consequences directly into the design, the amount of subjective judgement can be drastically reduced.

This chapter is not a critical review of statistical applications but a broad survey over those aspects of design to which statistical methods have potential use. Many important aspects have been omitted or treated very cursorily and the references to published work are by no means exhaustive. More emphasis is given to statistics (i.e. analysis of data) than to probability theory (i.e. response to random phenomena).

Some preliminary treatment of mathematical statistics must be given but this treatment is not detailed. For further information on such matters the standard reference is Kendall and Stuart (1958, 1961, 1966).

3.2   STATISTICAL CONCEPTS

3.2.1   Probability

Most engineers are familiar with the limiting frequency concept of probability, in which if an event A occurs on n occasions out of N the probability $P(A)$ of the event is taken as the limit

$$P(A) = \lim_{N \to \infty} (n/N) \quad ; \quad 0 \leqslant P(A) \leqslant 1$$

While this concept is satisfactory for many purposes it is not suitable when the probability of an unrepeatable event is being considered, as in most decision problems. Here, probability is best regarded as a measure of the strength of belief or degree of credibility associated with a particular outcome, (Jeffreys (1961) and Savage (1954)).

Conditional Probability.   If the probability of event A is associated with the occurrence of another event B, the probability of A *given that* the event B has occurred is the conditional probability $P(A|B)$. All probabilities are in fact conditional on something, but for brevity need not always be written in the conditional form.

The *joint probability* of both A and B occurring together $P(AB)$ is obviously

$$P(AB) = P(A|B) . P(B) = P(B|A) . P(A) \tag{3.1}$$

and if A and B occur quite independently of each other then $P(A|B) = P(A)$ and

$$P(AB) = P(A) . P(B) \tag{3.1a}$$

The *union probability* of either A or B or both A and B occurring is

$$P(A \cup B) = P(A) + P(B) - P(AB) \tag{3.2}$$

and if the two events are *mutually exclusive* and cannot possibly occur together, $P(AB) = 0$ and

$$P(A \cup B) = P(A + B) = P(A) + P(B) \tag{3.2a}$$

For a series of n independent events A_i $(i = 1$ to $n)$

$$P(A_1 A_2 \ldots A_i \ldots A_n) = \prod_{i=1}^{n} P(A_i) \tag{3.2b}$$

and for n mutually exclusive events

$$P(A_1 + A_2 + \ldots + A_i + \ldots + A_n) = \sum_{i=1}^{n} P(A_i) \tag{3.2c}$$

Bayes' theorem.   If an observed event A could have been due to one of n different possible causes B_i, mutually exclusive and exhaustive so that one of the B_i is certain to be the true cause,

$$\sum_{i=1}^{n} P(B_i) = 1$$

then

$$P(A) = \sum_{i=1}^{n} P(B_i) \cdot P(A|B_i)$$

$$P(AB_i) = P(A) \cdot P(B_i|A) = P(B_i) \cdot P(A|B_i)$$

or

$$P(B_i|A) = \frac{P(A|B_i) \cdot P(B_i)}{\sum_{i=1}^{n} P(A|B_i) \cdot P(B_i)} \tag{3.3}$$

This is Bayes' Theorem, which relates the probability of the cause B_i, given the event A is actually observed, to the probability of the event A given the cause B_i.

$P(B_i)$ is the *prior* probability of B_i (before observation of A).
$P(A|B_i)$ is the *likelihood* of the event, given B_i.
$P(B_i|A)$ is the *posterior probability* of B_i (after observation of A).

3.2.2 Random variables

All the factors appearing in a design are variable quantities whose values must be specified, estimated, or controlled within certain limits for the design to be successful. Exact values can rarely if ever be postulated in advance and there is inevitably a degree of uncertainty associated with the value actually realised. If the whole possible range of values is thought of as some hypothetical 'population', the actual values occurring in any one case may be thought of as a random sample obtained from this population.

The population is characterised by a *probability density function* (pdf) such that the probability of the variable X lying within the range

$$(x - \tfrac{1}{2}\,\mathrm{d}x) \leqslant X \leqslant (x + \tfrac{1}{2}\,\mathrm{d}x)$$

is

$$\mathrm{Prob}\{(x - \tfrac{1}{2}\,\mathrm{d}x) \leqslant X \leqslant (x + \tfrac{1}{2}\,\mathrm{d}x)\} = g(x|\theta_j)\mathrm{d}x$$

where $g(x|\theta_j)$ is the pdf, depending on k parameters θ_j, $j = 1$ to k.

The *cumulative distribution function* (cdf) is the probability that the variable is less than a particular value x and will be written as

$$P(x) \equiv \mathrm{Prob}(X \leqslant x) = \int_{-\infty}^{x} g(x|\theta_j)\mathrm{d}x = \int \mathrm{d}P(x)$$

The variable X is called a *random variable* and is defined by its pdf or cdf.

The inverse function or *quantile* x_p is the value of the random variable corresponding to a probability p,

$$P(x_p) = p$$

For two random variables X and Y the joint probability density function $h(x, y|\theta_j)$ and joint cumulative distribution function are related by

$$P(x, y) \equiv \mathrm{Prob}(X \leqslant x, Y \leqslant y) = \int_{-\infty}^{y} \int_{-\infty}^{x} h(x, y|\theta_j)\mathrm{d}x \cdot \mathrm{d}y$$

If one variate has a fixed value $Y = y_1$ say, then $P(x|y_1)$ is the conditional distribution of X and is given by

$$P(x|y_1) = \frac{\int_{-\infty}^{x} h(x, y_1|\theta_j)dx}{\int_{-\infty}^{\infty} h(x, y_1|\theta_j)dx} = \int_{-\infty}^{x} g(x|y_1, \theta_j)dx$$

The marginal distributions of X and Y are

$$P(x) = \int_{-\infty}^{x} dx \int_{-\infty}^{\infty} h(x, y|\theta_j)dy = \int_{-\infty}^{x} g_1(x)dx$$

$$P(y) = \int_{-\infty}^{y} dy \int_{-\infty}^{\infty} h(x, y|\theta_j)dx = \int_{-\infty}^{y} g_2(y)dy$$

If X and Y are independent then $P(x, y) = P(x).P(y)$ which implies that

$$h(x, y) = g_1(x).g_2(y)$$

otherwise

$$h(x, y) = g_1(x).g(y|x) = g_2(y).g(x|y)$$

Moments and expectations. The *expected* value of a random variable is the average value or the mean, defined by

$$E(X) = \mu = \int_{-\infty}^{\infty} x.g(x|\theta_j)dx = \int_{0}^{1} x.dP$$

and the mean is the first moment of the distribution about the origin. Similarly, the expected value of X^r would be the the r-th moment about the origin

$$E(X^r) = \mu_r' = \int_{0}^{1} x^r.dP \tag{3.4a}$$

Central moments, calculated about the mean, are of more use than the moments about the origin

$$E\{X - E(X)\}^r = \mu_r = \int_{0}^{1} (x - \mu)^r.dP \tag{3.4b}$$

The second central moment $\mu_2 = E\{X - E(X)\}^2$ is particularly useful and is called the *variance*, and will be given the special symbols $V(X)$ or σ^2, with $\sigma = \{V(X)\}^{\frac{1}{2}}$ the *standard deviation*. The mean and variance are analogous to the centroid and second moment of area of a section, with the standard deviation analogous to the radius of gyration. Means and variances are of great practical value in measuring the location and spread of a distribution, and are the most important of the parameters θ_j of the pdf.

A *standardised random variable* U can always be defined in the dimensionless form

$$U = (X - \mu)/\sigma$$

and U has zero mean and unit variance. Tabulated probabilities and quantiles of a distribution are almost always given for the standardised form. The *coefficient of variation* $CV(X) = C_x = \sigma/\mu$ is another useful dimensionless factor. In terms of the standardised random variable the actual variable is

$$X = \mu(1 + C_x \cdot U)$$

The standardised third and fourth central moments give some information on the shape of the pdf and are often called shape parameters

$$\sqrt{\beta_1} = E(U^3) = \mu_3/\sigma^3$$

$$\beta_2 = E(U^4) = \mu_4/\sigma^4$$

If the distribution is symmetrical about the mean, then $\sqrt{\beta_1} = 0$. For a non-symmetrical distribution $\sqrt{\beta_1}$ can be positive or negative, depending on the sign of μ_3, and is a measure of the *skewness* (*see Figure 3.1*). A rough

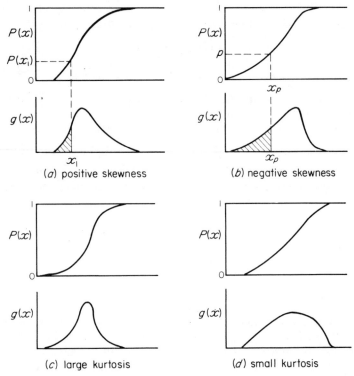

Figure 3.1   *Distribution functions*

indication of how the distribution is concentrated about the mean is given by β_2, small values implying a flatter curve for the pdf than large values (*see Figure 3.1*), and β_2 is a measure of the *kurtosis*.

For a bi-variate distribution of pdf $h(x, y)$, mixed central moments of the form

$$E[\{X - E(X)\}^r \cdot \{Y - E(Y)\}^s] = \mu_{rs} = \int_{-\infty}^{\infty} \int_{-\infty}^{\infty} (x - \mu_x)^r \cdot (y - \mu_y)^s \cdot h(x, y) \mathrm{d}x \cdot \mathrm{d}y$$

can be calculated. The first order mixed moment μ_{11} is of great importance, and is called the *covariance* of X and Y, $\text{Cov}(X, Y) = \sigma_{xy}$. If X and Y are independent then $\text{Cov}(X, Y) = 0$.

For multi-variate distributions the moments and expectations are defined in a similar fashion to that above.

3.2.3   Estimation

The parameters θ_j defining the pdf must be estimated from statistics $t_{j,n}$ obtained from a sample of n values $x_1, x_2, \ldots, x_i, \ldots, x_n$. Different samples would give different estimates, and consequently these estimated parameters are themselves random variables with their own means, variances, and pdf's.

The 'best' estimators are consistent, $\lim\limits_{n \to \infty} P(|t_n - \theta| < \varepsilon) \to 1$; are unbiased, $E(t_n - \theta) = 0$; have minimum variance $V(t_n)$; and are easy to calculate. General methods of estimation (see Kendall and Stuart (1961)) are as follows.

Bayesian estimation.   In equation 3.3 if B_i is taken as the wanted parameter θ_j and A as the sample data x_i, then the best estimate of θ_j would be that which maximised the posterior probability $P(\theta_j | x_i)$.

Maximum likelihood.   With equal prior probabilities $P(\theta_j)$ maximising the posterior probability reduces to maximising the likelihood $P(x_i | \theta_j)$.

Method of moments.   Expressing the wanted parameters in terms of the population moments μ_r and substituting the sample moments m_r will give enough equations to find θ_j.

Least squares.   If the parameter equations can be written in the form $f(x | \theta_j) = 0$ then estimates can be obtained by minimising

$$\sum_{i=1}^{n} \{f(x_i | \theta_j)\}^2$$

from

$$\frac{\partial}{\partial \theta_j} \left[\sum_{i=1}^{n} \{f(x_i | \theta_j)\}^2 \right] = 0$$

Quantiles.   The theoretical quantiles corresponding to a probability p will be functions of the parameters. Equating k observed quantiles from the sample to the population quantiles gives enough equations to solve for the parameters.

Order statistics.   Rearranging the sample values in ascending order $x_{(1)} < x_{(2)} < \ldots < x_{(n)}$ the expected values of $x_{(i)}$ will depend on the parameters. Linear combinations of these order statistics $x_{(i)}$ are equated to the corresponding expectations.

Bayesian, maximum likelihood, and least squares methods are more efficient than the others, but are often very complicated in practice. For medium sized samples (n from about 20 to about 100) the method of moments is generally satisfactory, although theoretically inefficient, while for large samples (n greater than about 100) the method of quantiles and for small samples (n less than about 20) the use of order statistics will often be quite satisfactory.

3.2.4 Inference

From the population pdf, the pdf of the estimated parameter $t_{j,n}$ can be determined and the cumulative distribution $P(t_{j,n})$ calculated. Consequently upper and lower *confidence limits* for $(t_{j,n} - \theta_j)$ could be determined from

$$\text{Prob}\{(t_{j,n} - \theta_j) \geqslant u_1\} = \alpha$$

$$\text{Prob}\{(t_{j,n} - \theta_j) \geqslant -u_2\} = 1 - \beta$$

or, equivalently, the unknown true value θ_j bracketed in a known range with associated probability $p = 1 - \gamma = 1 - (\alpha + \beta)$

$$\theta_{min} = (t_{j,n} - u_1) \leqslant \theta_j \leqslant (t_{j,n} + u_2) = \theta_{max}$$

These upper and lower confidence limits θ_{min} and θ_{max} will depend on the value of γ and on the number of *degrees of freedom* associated with the estimate $t_{j,n}$. Roughly speaking, if k parameters must be estimated before $P(t_{j,n})$ can be calculated then the number of degrees of freedom $v = (n - k)$.

A value of $(1 - \gamma)$ must be chosen and usually this is taken as a quite arbitrary value of 0.9, 0.95, or 0.99. On a relative frequency definition of probability the interpretation would be that θ_j would fall within the two limits in a proportion of $(1 - \gamma)$ *repeated* samplings all of size n from the population. In practice it is very unlikely that any repeated samplings would ever be performed and $(1 - \gamma)$ is best interpreted as a degree of belief that θ_j does lie between the two limits.

In many cases only one-sided confidence limits are required, to give either $\theta_j \leqslant \theta_{max}$ or $\theta_j \geqslant \theta_{min}$, in which case the probability $(1 - \gamma)$ is either $(1 - \alpha)$ or $(1 - \beta)$. For the two-sided limits where $(1 - \gamma) = 1 - (\alpha + \beta)$ it is common practice to take $\alpha = \beta = \frac{1}{2}\gamma$. Thus, for 0.95 probability confidence limits $\alpha = \beta = 0.025$ for the two-sided case and 0.05 for the one-sided case.

When confidence limits are applied simultaneously to k parameters calculation of the joint pdf of the k variates may be very difficult. An approximation to the joint probability $(1 - \gamma)$ for such cases may be obtained from equation 3.2b

$$(1 - \gamma) = 1 - \sum_{j=1}^{k} (\alpha_j + \beta_j)$$

with α_j and β_j the probabilities for the individual limits, generally taken as equal. Thus, for $(1 - \gamma) = 0.95$ and 10 simultaneous two-sided limits, the individual probabilities for each parameter would be $\alpha_j = \beta_j = 0.0025$.

3.3 DISTRIBUTION FUNCTIONS

Although any positive function $g(x) \geqslant 0$ is a probability density function, provided that $\int_{-\infty}^{\infty} g(x) dx = 1$, certain particular standard distributions are of outstanding importance.

3.3.1   Univariate continuous distributions

Some of the most useful standard forms are members of the Pearson family of distributions defined by

$$\frac{d}{dx}\{\log . g(x)\} = \frac{x - c_0}{c_1 + c_2 x + c_3 x^2} \quad ; \quad c_i = \text{constant}$$

Three main types of distribution can be distinguished on the basis of the possible range of x (depending on whether the roots of the denominator are real or imaginary) as follows

Main Type 1   (Pearson's types I and II)
$$a \leqslant x \leqslant b \qquad\qquad \text{Bounded}$$

Main Type 2   (Pearson's type VI)
$$a \leqslant x < \infty \qquad\qquad \text{Semi-bounded}$$
$$-\infty < x \leqslant b$$

Main Type 3   (Pearson's type IV)
$$-\infty < x < \infty \qquad\qquad \text{Unbounded}$$

The parameters in the pdf's are determined by the Method of Moments (Elderton and Johnson (1969)) and for most practical cases it is sufficient to calculate the sample mean $\bar{x}$ and variance s^2 together with skewness $\sqrt{b_1}$ and kurtosis b_2 from unbiased estimates of the moments.

$$m_r = \frac{1}{n}\sum_{i=1}^{n}(x_i - \bar{x})^r \quad ; \quad \bar{x} = \frac{1}{n}\sum_{i=1}^{n}x_i$$

$$s^2 = \mu_2 = \frac{n}{n-1}m_2$$

$$\mu_3 = \frac{n^2}{(n-1)(n-2)}m_3$$

$$\mu_4 = \frac{n^2}{(n-1)(n-2)(n-3)}\{(n+1)m_4 - 3(n-1)m_2^2\} + 3s^4$$

$$\sqrt{b_1} = \mu_3/s^3 \quad ; \quad b_2 = \mu_4/s^4$$

The cumulative distribution function is then found from the tables of standardised quantiles $u_p = (x - \bar{x})/s$ given by Johnson et al. (1963) for given values of $\sqrt{\beta_1}$ and β_2.

Since the Method of Moments is theoretically inefficient the variances of $\sqrt{b_1}$ and b_2 can be quite large, as shown by the large-sample standard deviations of Figure 3.2, which gives $\sqrt{n} . \sigma_{\sqrt{b_1}}$ and $\sqrt{n} . \sigma_{b_2}$ as functions of $\sqrt{\beta_1}$ and β_2, but the simplicity of the Method of Moments in comparison with the more efficient Bayesian or Maximum Likelihood Methods outweighs the theoretical disadvantages.

Two special limiting forms of Main Type 1 and Main Type 3 distributions are the Uniform or Rectangular, and the Normal or Gaussian distributions.

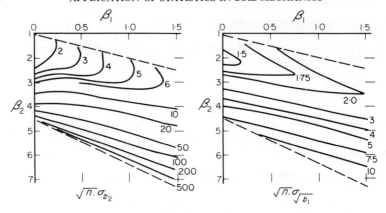

Figure 3.2   Standard deviations of b_2 and $\sqrt{b_1}$

Uniform distribution.
$$g(x) = 1/(b-a) \qquad a \leqslant x \leqslant b$$
$$\mu = \tfrac{1}{2}(a+b)$$
$$\sigma^2 = \tfrac{1}{12}(b-a)^2$$
$$\sqrt{\beta_1} = 0$$
$$\beta_2 = 1.8$$

The parameters a and b are best estimated from the largest and smallest sample values $x_{(n)}$ and $x_{(1)}$ as

$$\tfrac{1}{2}(a+b) = \tfrac{1}{2}\{x_{(n)}+x_{(1)}\}$$

$$(b-a) = \frac{n+1}{n-1}\{x_{(n)}-x_{(1)}\}$$

Normal distribution.   In standardised form with $u = (x-\mu)/\sigma$
$$g(u) = (2\pi)^{-\frac{1}{2}}.\exp(-\tfrac{1}{2}u^2) \qquad -\infty < u < \infty$$
$$\sqrt{\beta_1} = 0$$
$$\beta_2 = 3$$

The mean and variance, the two parameters, are estimated from

$$\mu = \bar{x} = \frac{1}{n}\sum x$$

$$\sigma^2 = s^2 = \frac{1}{n-1}\sum (x-\bar{x})^2$$

This distribution is of fundamental importance in the theory of statistics, principally because of the Central Limit Theorem (see Section 3.3.4), but cannot be expected to fit all the random variables encountered in practice.

Log-normal distribution.   If z is a standardised Normal variate, the Log-normal distribution of x is defined by

$$z = a+b.\log(u-d) \qquad d \leqslant u < \infty$$

with $u = (x-\bar{x})/\sigma$ as before.

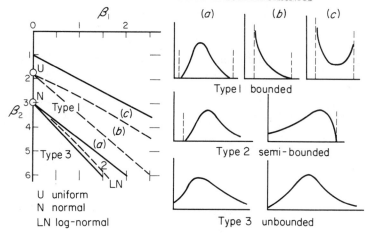

Figure 3.3   Pearson distribution functions

The various parameters are best estimated by transforming from x to $\log . x$ and then treating $\log . x$ as a Normal variate.

The range of validity of the three main types of Pearson Curves, together with the Log-normal curve, are shown in *Figure 3.3* in terms of the skewness and kurtosis parameters, together with typical shapes of the pdf's.

3.3.2   Bi-variate and multi-variate distributions

When two random variables X and Y are involved, the joint distribution pdf $h(x, y \mid \theta_j)$ must be estimated from the n pairs of results x_i, y_i in the sample. Marginal distributions can always be fitted as Pearson Curves but, unless X and Y are independent, these marginal distributions are insufficient to define $h(x, y)$. Generally the best that can be done with a reasonably large sample (n of the order of 50 to 200) is to assume a particular distribution, such as the Bi-variate Normal or Bi-variate Uniform, and to hope that it will be adequate.

Bi-variate uniform distribution.   The variables X and Y will be bounded by an area $A = f(x, y)$ and the pdf is

$$h(x, y) = 1/A \qquad |x, y| \leqslant A$$
$$h(x, y) = 0 \qquad |x, y| > A$$

Bi-variate normal or bi-normal distribution.   In standardised form with u, v given by $u = (x - \mu_x)/\sigma_x$, $v = (y - u_y)/\sigma_y$, the pdf of u, v is

$$h(u, v \mid \rho) = \{2\pi(1 - \rho^2)^{\frac{1}{2}}\}^{-1} . \exp\left\{ -\frac{1}{2(1 - \rho^2)} . (u^2 - 2\rho uv + v^2) \right\}$$

where ρ, the *correlation coefficient* is defined by

$$\rho = \frac{\text{cov}(X, Y)}{\{V(X) . V(Y)\}^{\frac{1}{2}}} = \frac{\sigma_{xy}}{\sigma_x . \sigma_y} \qquad ; \qquad -1 \leqslant \rho \leqslant 1$$

Distributions of directions. A very common case of two variates is a distribution of unit vectors or directions, such as the inclinations and bearings of joints, fissures, or other discontinuities in a soil volume. The pole of a joint-plane with dip angle θ' and bearing angle φ' can be regarded as a vector of unit length, and a set of joint-planes would be a set of vectors (θ'_i, φ'_i) which can be plotted on the surface of a unit sphere. Two frequently occurring cases are for the vectors to cluster around a modal vector (θ_0, φ_0) or to be spread in a girdle at 90° to the anti-mode (θ_0, φ_0). The coordinates of mode or anti-mode are found from the direction cosines $\bar{x}, \bar{y}, \bar{z}$ of the resultant vector $\boldsymbol{R}$.

$$\bar{x} = \frac{1}{n}\sum_i^n x_i = \frac{1}{n}\sum \sin\theta'_i . \cos\varphi'_i$$

$$\bar{y} = \frac{1}{n}\sum_i^n y_i = \frac{1}{n}\sum \sin\theta'_i . \sin\varphi'_i$$

$$\bar{z} = \frac{1}{n}\sum_i^n z_i = \frac{1}{n}\sum \cos\theta'_i$$

$$\boldsymbol{R}^2 = \sum_i^n x_i^2 + \sum_i^n y_i^2 + \sum_i^n z_i^2$$

$$\cos\theta_0 = \bar{z}$$

$$\tan\varphi_0 = \bar{y}/\bar{x}$$

Measuring dip θ and bearing φ from the mode, the pdf for a clustered distribution, analogous to the Bi-normal with $\rho = 0$ (symmetrical about the mode), is Fisher's Distribution (Fisher (1953))

$$h(\theta, \varphi \mid \lambda) = \frac{\lambda}{4\pi \sinh\lambda} . \exp(\lambda\cos\theta) \qquad \begin{array}{l} 0 \leqslant \theta \leqslant \pi \\ 0 \leqslant \varphi \leqslant 2\pi \end{array}$$

with $$P(\theta) = \frac{\lambda}{2\sinh\lambda}\{\exp(\lambda) - \exp(\lambda\cos\theta)\}$$

and an approximate estimate of λ is

$$\lambda = \frac{n-1}{n-\boldsymbol{R}}$$

For a girdle distribution, symmetrical about the anti-mode, a suitable form is Selby's Distribution (Selby (1964))

$$h(\theta, \varphi \mid \kappa) = \frac{1}{4\pi} . \frac{\kappa}{1 - \exp(-\kappa)} . \exp(-\kappa |\cos\theta|) \qquad \begin{array}{l} 0 \leqslant \theta \leqslant \pi \\ 0 \leqslant \varphi \leqslant 2\pi \end{array}$$

$$P(\theta) = \frac{\exp(-\kappa |\cos\theta|) - \exp(-\kappa)}{1 - \exp(-\kappa)}$$

with κ estimated from

$$\kappa = \frac{n}{\sum |\cos\theta_i|}$$

3.3.3   Goodness of fit

The best distribution function to choose is that which gives an adequate fit to the results with the least number of parameters. It is always advisable to check the fit of the assumed distribution after estimating the parameters.

The chi-squared test.   In the standard test of goodness of fit the n sample values are grouped into k classes x_0 to x_1, x_1 to $x_2, \ldots, x_{i-1}$ to $x_i, \ldots, x_{k-1}$ to x_k, and the number of observations n_i falling into the i-th class counted. From the assumed cdf the expected number e_i in the i-th class can be calculated from

$$e_i = n\{P(x_i) - P(x_{i-1})\}$$

and χ^2 calculated as

$$\chi^2 = \sum_{i=1}^{k} \frac{(n_i - e_i)^2}{e_i}$$

If there are r parameters to be estimated before e_i can be calculated then χ^2 has $\nu = k - r$ degrees of freedom, and if χ^2 is less than the critical value $(\chi^2)_\nu^\gamma$ which could arise by chance with a probability γ, $\text{Prob}(\chi^2 \geqslant \chi^2_\nu) = \gamma$, then there is no significant difference between the assumed distribution and the observations, at a significance level γ.

Table 3.1   LIQUIDITY INDEX OF MARINE CLAY

1.026	0.936	0.918	0.885	0.870	0.854	0.834	0.821	0.800	0.753
0.989	0.934	0.909	0.882	0.870	0.854	0.832	0.816	0.787	0.751
0.981	0.934	0.909	0.882	0.870	0.850	0.831	0.815	0.781	0.749
0.969	0.932	0.904	0.879	0.869	0.848	0.831	0.814	0.779	0.745
0.966	0.932	0.900	0.879	0.866	0.846	0.830	0.814	0.779	0.731
0.954	0.932	0.897	0.878	0.865	0.846	0.830	0.814	0.769	0.731
0.950	0.929	0.894	0.878	0.862	0.844	0.829	0.811	0.767	0.729
0.950	0.929	0.893	0.875	0.861	0.840	0.826	0.810	0.766	0.724
0.949	0.926	0.889	0.874	0.861	0.837	0.826	0.809	0.766	0.722
0.942	0.925	0.889	0.874	0.860	0.836	0.825	0.806	0.760	0.719
0.939	0.924	0.886	0.874	0.857	0.834	0.824	0.805	0.758	0.709
0.938	0.922	0.885	0.872	0.856	0.834	0.822	0.805	0.757	0.701

Table 3.1 gives 120 results of liquidity index for a normally consolidated marine clay (Lumb (1966)), which will be fitted to a Normal distribution. The mean and standard deviation were estimated to be 0.852 and 0.0690. Taking $k = 20$ classes with equal class probabilities $P(x_i) - P(x_{i-1}) = 1/k = 0.05$ the class boundaries and observed values n_i are shown in *Table 3.2*. With equal expected values $e_i = n/k$, χ^2 is calculated from

$$\chi^2 = \frac{k}{n} \sum_{i=1}^{k} (n_i)^2 - n$$

and is 24.4. Three parameters n, μ, σ^2, are required so $\nu = k - r = 20 - 3 = 17$. With 17 degrees of freedom the probability of χ^2 being greater than 24.4 is 10.9%. This is sufficiently large a probability for the deviations between observed and expected numbers to be attributed to chance sampling variations, and it can be safely assumed that the liquidity index does follow the Normal distribution.

Table 3.2   CLASS BOUNDARIES AND OBSERVED NUMBERS FOR LIQUID LIMIT

P	u	x	n_i	P	u	x	n_i
0.00–0.05			8	0.50–0.55			6
	−1.645	0·738			0.126	0.861	
0.05–0.10			7	0.55–0.60			$5\frac{1}{2}$
	−1.282	0.763			0.253	0.870	
0.10–0.15			$6\frac{1}{2}$	0.60–0.65			$9\frac{1}{2}$
	−1.036	0.781			0.385	0.879	
0.15–0.20			$1\frac{1}{2}$	0.65–0.70			6
	−0.842	0.794			0.524	0.888	
0.20–0.25			$3\frac{1}{2}$	0.70–0.75			5
	−0.675	0.806			0.675	0.898	
0.25–0.30			8	0.75–0.80			4
	−0.524	0.816			0.842	0.910	
0.30–0.35			$4\frac{1}{2}$	0.80–0.85			2
	−0.385	0.825			1.036	0.923	
0.35–0.40			$10\frac{1}{2}$	0.85–0.90			13
	−0.253	0.834			1.282	0.941	
0.40–0.45			$4\frac{1}{2}$	0.90–0.95			$5\frac{1}{2}$
	−0.126	0.843			1.645	0.966	
0.45–0.50			5	0.95–1.0			$4\frac{1}{2}$
	0.000	0.852					

Kolmogorov-Smirnov test.   An alternative test is to measure the maximum absolute difference D_n between the sample cumulative distribution $F_n(x)$ and the assumed population cdf $P(x)$.

$$D_n = \sup |F_n(x) - P(x)|$$

If n is larger than about 40 the critical values D_n^γ such that $\text{Prob}(D \geqslant D_n^\gamma) = \gamma$ are

γ	0.20	0.10	0.05	0.01
$\sqrt{n}.\,D_n^\gamma$	1.07	1.22	1.36	1.63

Figure 3.4 shows the cdf of the liquidity index results. The maximum absolute difference at a liquidity index of 0.923 is $(0.850 - 0.808) = 0.042$, and this is less than the critical value at 20% significance, $1.07/(120)^{\frac{1}{2}} = 0.098$. Again, it can be safely assumed that the Normal distribution is satisfactory.

Skewness and Kurtosis tests.   A third test is to compare the sample and theoretical values of skewness and kurtosis. For the liquidity index results the sample skewness $\sqrt{b_1} = -0.10$, and the sample kurtosis $b_2 = 2.62$. For the Normal distribution there is a 90% probability (Pearson and Hartley (1966)) that the sample values for a size $n = 120$ would lie between

$$-0.35 < \sqrt{b_1} < 0.35$$
$$2.40 < b_2 < 3.71$$

and since the sample results are well within these limits, again the conclusion is that the Normal distribution is satisfactory.

Small sample tests.   If the size of the sample is less than about 30, the three tests above have little discriminating power. A useful small-sample

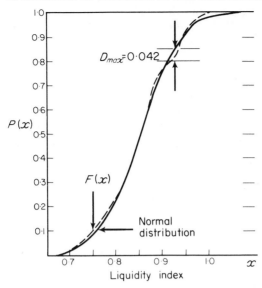

Figure 3.4 Sample and population distributions for liquidity index

test, based on the order statistics, is the W-test of Shapiro and Wilk (1965) for the Normal distribution.

$$W_n = \frac{b^2}{(n-1)s^2} = \frac{\left[\sum\limits_{i=1}^{n} \{a_i . x_{(i)}\}\right]^2}{\sum\limits_{i=1}^{n} (x_i - \bar{x})^2}$$

where the *weights* a_i depend on n and are given by Shapiro and Wilk.

Table 3.3 shows ordered values of $t = \tan \varphi$, the tangent of the angle of shearing resistance of undisturbed cohesive till, from 22 drained shear box tests, together with the weights a_i.

$$b = 0.459(0.611 - 0.306) + \ldots + 0.012(0.480 - 0.480)$$
$$= 0.3450$$
$$(n-1)s^2 = 0.1216$$
$$W_{22} = (0.3450)^2/(0.1216)$$
$$= 0.98$$

From Shapiro and Wilk's tables a W-value of 0.98 could arise by chance with a probability of 90%, for a sample of size 22 from the Normal distribution, and hence Normality can be safely assumed.

Quantile plots. A graphical display of the sample distribution is always useful when assessing goodness of fit, and the most satisfactory type of display is the Quantile Plot (Wilk and Gnanadesikan (1968)) in which the sample quantiles x_p are plotted against the corresponding assumed population quantiles ξ_p for the same probability p. For perfect agreement the results should all fall on a line inclined at 45°, and deviations from this

Table 3.3 ORDERED VALUES OF TAN ϕ FOR COHESIVE TILL, WITH NORMAL WEIGHTS

i	t	a_i	i	t	a_i
1	0.306	−0.459	12	0.480	0.012
2	0.344	−0.316	13	0.483	0.037
3	0.368	−0.257	14	0.494	0.062
4	0.390	−0.213	15	0.499	0.088
5	0.409	−0.176	16	0.500	0.115
6	0.416	−0.144	17	0.528	0.144
7	0.432	−0.115	18	0.533	0.176
8	0.440	−0.088	19	0.544	0.213
9	0.475	−0.062	20	0.559	0.257
10	0.479	−0.037	21	0.568	0.316
11	0.480	−0.012	22	0.611	0.459

diagonal indicate discrepancies between observed and assumed distributions, or lack of fit.

With small samples (n less than about 30) all the observed values can be plotted, but there is a certain ambiguity in the choice of plotting position of the ordered values $x_{(i)}$. If the probability p_i for the i-th value were taken as either (i/n) or $(i-1)/n$ then one or the other of the two extremes $x_{(1)}$ or $x_{(n)}$ would be represented by $p = 0$ or $p = 1$, which is obviously wrong. Either the average $(i-\frac{1}{2})/n$ or the expected value of the order statistic $i/(n+1)$ is more satisfactory, and this latter value will be used here. *Figure 3.5* shows the quantile plot for the liquidity index of *Table 3.1* and the plot for the tangent of angle of shearing resistance of *Table 3.3*, plotted against Normal quantiles.

Multi-variate distributions. For bi-variate or multi-variate distributions there are no analogues of the Kolmogorov–Smirnov, Shape Parameter, W-Test, or Quantile Plot methods, and no alternative to the Chi-squared test.

The k classes of the one-dimensional grouping are replaced by k contiguous regions or cells, and obviously these cells can be chosen in many ways. If a

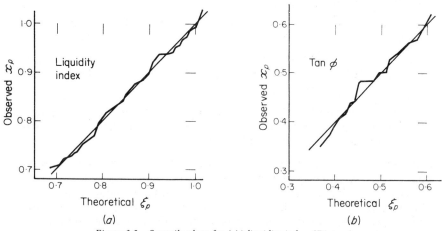

Figure 3.5   Quantile plots for (A) liquidity index (B) tan ϕ

fit to the Bi-Normal distribution were being tested, an elliptical region defined by

$$\frac{1}{1-\rho^2}\{u^2-2\rho uv+v^2\} = 2c^2$$

would contain a proportion $p = 1-\exp(-c^2)$ of the total points, and a series of nested ellipses can be constructed so that successive pairs contain equal proportions $(1/k_1)$. Alternatively, the u, v region could be divided into k_2 sectors such that each sector contains equal proportions $(1/k_2)$, by lines through the origin at angles β_j such that

$$\tan \beta_j = \frac{1+m.\tan(2\pi\rho_j)}{1-m.\tan(2\pi\rho_j)}$$

where

$$m^2 = (1-\rho)/(1+\rho)$$

Combining ellipses and sectors the u, v region can be divided into $k = k_1.k_2$ quadrats, each of which would contain an expected number $(n/k_1.k_2)$ of points, as shown in Figure 3.6.

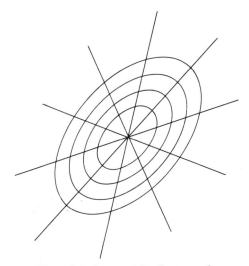

Figure 3.6   Bi-normal distribution quadrats

Figure 3.7 shows a scatter diagram of standardised variate pairs u (from the drained cohesion) and v (from the tangent of the drained angle of shearing resistance) for 77 triaxial compression tests on undisturbed residual soils. Dividing the region into 13 equi-probable quadrats, on the assumption that u and v are uncorrelated Bi-Normal (i.e. that the correlation coefficient is zero), the number of observed points n_i is given in Table 3.4. The expected number per quadrat is 5.92 and hence $\chi^2 = 11.5$. Five parameters (two means, two variances, and the total number) are required to calculate the expected values, so $v = k-r = 8$. For 8 degrees of freedom the probability of χ^2 being at least equal to 11.5 is 17.5%, and the assumption of uncorrelated Bi-Normality is quite reasonable.

Table 3.4 OBSERVED NUMBERS OF STANDARDISED COHESION AND TAN ϕ FOR RESIDUAL SOILS

Cell number	1	2	3	4	5	6	7
Observed number	6	4	7	7	4	6	6

Cell number	8	9	10	11	12	13	
Observed number	12	8	7	4	2	3	

Multi-modal distributions. All the distributions described above have a single unique maximum to the pdf, that is they are uni-modal, but observed distributions are commonly found to have two or more modes, as shown in *Figure 3.8.* Theoretical bi-modal distributions can obviously be produced by combining any two uni-modal distributions of pdf's $g_1(x)$ and $g_2(x)$ through

$$g(x) = (1-k)g_1(x) + k \cdot g_2(x)$$

if the two means are unequal.

Decomposition of an observed bi-modal distribution into two constituent parts $g_1(x), g_2(x)$ is extremely difficult, if efficient methods are used. Even for

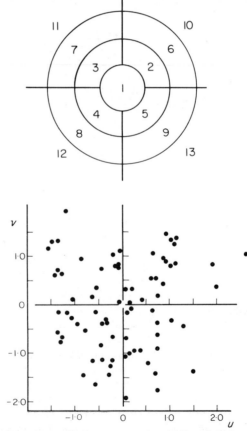

Figure 3.7   Bi-variate standardised variables and quadrats

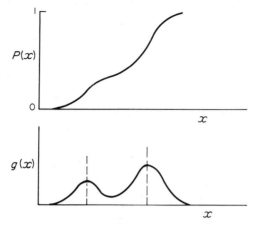

Figure 3.8   Multi-modal distribution

the simplest case of two Normal components, the general solution involves the sixth sample moment m_6, which is very sensitive to sampling variations. Iterative methods are available (Hasselblad (1966)) but unless the sample is very large (n greater than about 400) the data hardly warrant a precise fit. It will usually be sufficient to use inefficient methods, assuming initial values for the means, variances, and relative proportions, refining the calculations by successive approximations until an acceptably low value of χ^2 is achieved.

Figure 3.9 shows a multi-modal distribution of joint-plane orientations in decomposed granite (Lumb (1971a)) on an equi-areal polar lower-hemi-spherical plot. There are three distinct clusters of joints, two with high dips and one with a low dip, and a satisfactory fit (Chi-squared probability 24%) is obtained by combining three Fisher distributions of parameters $(\theta_0, \varphi_0, \lambda)$ in proportions k as follows.

k	$\theta_0{}^\circ$	$\varphi_0{}^\circ$	λ
0.5	90	16	7
0.3	79	270	35
0.2	16	325	7

The theoretical probability contours are also shown on Figure 3.9.

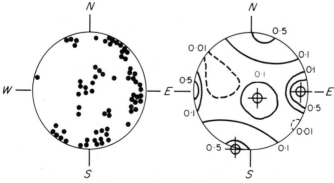

Figure 3.9   Distribution of joint-plane orientations

As an alternative to decomposition, the sample distribution itself could be used as an estimate of the population distribution and, for a one-dimensional multi-modal distribution, confidence limits can be calculated by inverting the Kolmogorov–Smirnov statistic. For a sample cdf $F_n(x)$ the population cdf $P(x)$ would be bracketed by D_n^γ so that

$$\text{Prob}\{[F_n(x)-D_n^\gamma] < P(x) < [F_n(x)+D_n^\gamma]\} = 1-\gamma$$

Unfortunately the corresponding confidence limits for the quantiles x_p are infinitely wide if p is less than D_n or greater than $(1-D_n)$ as can be seen from

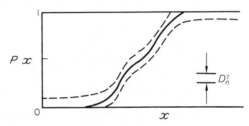

Figure 3.10   Kolmogorov-Smirnov confidence limits

Figure 3.10 and, since these extreme quantiles are what are commonly required in practice, the confidence limits are of little value unless n is very large.

3.3.4   Functions of random variables

The response y of the soil will be given by some design equation $y = f(x_i)$, where the x_i are now n different random variables $x_1, x_2, \ldots, x_i, \ldots, x_n$, whose pdf's have been estimated. Since the x_i are random variables the response y will also be a random variable and the pdf of y must be estimated.

If $f(x)$ and the pdf's $g_i(x_i)$ are of known analytical form there are exact methods for calculating the pdf $g(y)$ but the analysis can be extremely complicated unless the form of $f(y)$ and $g_i(x_i)$ is simple.

For two independent variables x_1 and x_2 the exact form of the pdf of a sum, $y = x_1+x_2$; a product, $z = x_1 . x_2$; or a quotient, $w = x_1/x_2$ is given by

$$g(y) = \int_{-\infty}^{\infty} g_1(y-x_2) . g_2(x_2) dx_2 \qquad (3.5a)$$

$$g(z) = \int_{-\infty}^{\infty} g_1(z/x_2) . |x_2|^{-1} . g_2(x_2) dx_2 \qquad (3.5b)$$

$$g(w) = \int_{-\infty}^{\infty} g_1(w . x_2) . |x_2| . g_2(x_2) dx_2 \qquad (3.5c)$$

and repeated use of these equations would eventually give the pdf of a function consisting of sums, products, and quotients, although evaluation of the integrals in a closed form is far from easy.

There are several approximate methods which are of more practical use than equation 3.5.

Normal and log-normal approximations. If y is a sum of n independent random variables, $y = \sum_{i=1}^{n} x_i$, it can be shown that the pdf of y is asymptotically Normal, that is

$$\lim_{n \to \infty} . g(y) = (2\pi\sigma^2)^{-\frac{1}{2}} . \exp\left\{-\frac{(y-\bar{y})^2}{2\sigma^2}\right\}$$

provided that the x_i have finite moments. This is the Central Limit Theorem, and can often be applied successfully even though n is small.

Similarly if y is a product of n independent variables, $y = \prod_{i=1}^{n} x_i$, then the pdf of $z = \log(y)$ is asymptotically Normal, or y is asymptotically Log-normal. Consequently for sums and products only the mean and variance of y or z need to be estimated.

For $y = \sum_{i=1}^{n} \beta_i x_i$ the mean and variance are simply

$$\bar{y} = \sum_{i=1}^{n} \beta_i \bar{x}_i \tag{3.6a}$$

$$V(y) = \sum_{i=1}^{n} \beta_i^2 . V(x_i) \tag{3.6b}$$

For $y = \prod_{i=1}^{n} (x_i)^{\alpha_i}$ some further assumptions are necessary. Writing

$$z = \log . y = \sum_{i}^{n} \alpha_i . w_i$$

with $w_i = \log . x_i$, then

$$\bar{z} = \sum_{i}^{n} \alpha_i . \bar{w}_i \tag{3.7a}$$

$$V(z) = \sum_{i}^{n} (\alpha_i)^2 . V(w_i) \tag{3.7b}$$

If $x_i = \bar{x}_i . (1 + C_i . u)$, where C_i is the coefficient of variation of x_i and *assuming* that u is Normally distributed, then

$$\bar{w}_i \simeq \log . \bar{x}_i - \tfrac{1}{2}(C_i)^2 - \tfrac{3}{4}(C_i)^4 \tag{3.7c}$$
$$V(w_i) \simeq (C_i)^2 + \tfrac{5}{2}(C_i)^4 \tag{3.7d}$$

Taylor's series approximation. If $y = f(x_i)$ is differentiable with respect to x_i, either analytically or numerically, it can be linearised by expanding in a Taylor's series about the mean $\bar{x}_i = (\bar{x}_1, \bar{x}_2, \ldots, \bar{x}_i, \ldots, \bar{x}_n)$. Writing $f(\bar{x})$ for the value of $f(x_i)$ at $\bar{x}_i$, f_i for $\partial f/\partial x_i$ evaluated at $\bar{x}_i$, f_{ij} for $\partial^2 f/\partial x_i \, \partial x_j$, $V(x_i)$

as σ_i^2, $\mathrm{Cov}(x_i, x_j)$ as σ_{ij}, and the third and fourth central moments of x_i as $\mu_3(i)$, $\mu_4(i)$, then approximately

$$\bar{y} = f(x) + \tfrac{1}{2}\sum_i^n f_{ii}\cdot\sigma_i^2 + \sum_{i<j}^n f_{ij}\cdot\sigma_{ij} \tag{3.8a}$$

$$V(y) = \sum_i^n (f_i)^2\cdot\sigma_i^2 + 2\sum_{i<j}^n f_i\cdot f_j\cdot\sigma_{ij} + \sum_i^n f_i\cdot f_{ii}\cdot\mu_3(i) \tag{3.8b}$$

$$\mu_3(y) = \sum_i^n (f_i)^3\cdot\mu_3(i) \tag{3.8c}$$

$$\mu_4(y) = \sum_i^n (f_i)^4\cdot\mu_4(i) + 6\sum_{i<j}^n (f_i)^2\cdot(f_j)^2\cdot(\sigma_i^2\cdot\sigma_j^2 + 2\sigma_{ij}^2) \tag{3.8d}$$

and from these equations the shape factors $\sqrt{\{\beta_1(y)\}}$, and $\beta_2(y)$ can be estimated and a Pearson Distribution fitted to give $P(y)$.

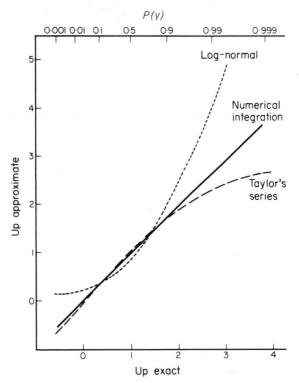

Figure 3.11   *Quantile plot for ratio of two normal distributions*

Numerical integration. When the form of $y = f(x_i)$ is not known analytically but can be found either numerically or experimentally, as is often the case in soil engineering, calculation of the differentials f_i, f_{ij}, may be difficult or inaccurate. The moments of y can still be found, however, quite accurately by numerical integration.

The expected value $E(h)$ of a function $h(x_i)$ can be written approximately (Evans (1967), (1972)) as

$$E(h) = C.h(\bar{x}) + \sum H_i . \left[\frac{h(a_i^+.\sigma_i)}{a_i^+} - \frac{h(a_i^-.\sigma_i)}{a_i^-} \right]$$

$$+ \sum P_{ij} . \left[\frac{h(a_i^+.\sigma_i; a_j^+.\sigma_j)}{a_i^+.a_j^+} - \frac{h(a_i^+.\sigma_i; a_j^-.\sigma_j)}{a_i^+.a_j^-} \right.$$

$$\left. - \frac{h(a_i^-.\sigma_i; a_j^+.\sigma_j)}{a_i^-.a_j^+} + \frac{h(a_i^-.\sigma_i; a_j^-.\sigma_j)}{a_i^-.a_j^-} \right] \tag{3.9a}$$

$C, H_i, P_{ij}, a_i^+, a_i^-$, are constants, $h(\bar{x})$ is the value of $h(x_i)$ at $\bar{x}_i$, $h(a_i^+.\sigma_i)$ the value at $\{\bar{x}_1,\dots,(\bar{x}_i+\sigma_i.a_i^+),\dots,\bar{x}_n\}$, and $h(a_i^+.\sigma_i; a_j^+.\sigma_j)$ the value at $\{\bar{x}_1,\dots,(\bar{x}_i+\sigma_i.a_i^+),\dots,(\bar{x}_j+\sigma_j.a_j^+),\dots,\bar{x}_n\}$, etc. The summations are over all possible combinations, thus there are n terms in the first sum and $\frac{1}{2}n(n-1)$ terms in the second sum.

For the mean or $E\{f(x_i)\}$ the function $h(x_i)$ is simply $f(x_i)$, while for the r-th central moment $h(x_i)$ is $[f(x_i)-E\{f(x_i)\}]^r$.

Temporarily writing the shape parameters of the i-th variate as $\sqrt{\{\beta_1(x_i)\}} = \delta_i$ and $\beta_2(x_i) = \Delta_i$ the required constants have been given by Evans as

where
$$a_i^+ = \tfrac{1}{2}\delta_i + R_i; \quad a_i^- = \tfrac{1}{2}\delta_i - R_i$$
$$C = 1 - S + \tfrac{1}{2}(S)^2 - \tfrac{1}{2}.T$$
$$H_i = (1 - S_i)/(2R_i)$$
$$P_{ij} = 1/(4R_i.R_j)$$
$$R_i = (\Delta_i - \tfrac{3}{4}.\delta_i^2)^{\frac{1}{2}}$$
$$S = \sum_i^n (\Delta_i - \delta_i^2)^{-1}$$
$$S_i = S - (\Delta_i - \delta_i^2)^{-1}$$
$$T = \sum_i^n (\Delta_i - \delta_i^2)^{-2}$$
$$\tag{3.9b}$$

In the derivation of equation 3.9 it is assumed that the n variables x_i are independent variables, that is $\mathrm{Cov}(x_i, x_j) = 0$.

If all variates are Normally distributed the constants of equation 3.9b simplify to

$$a_i^+ = -a_i^- = \sqrt{3}$$
$$C = 1 + n(n-7)/18$$
$$H_i = -(n-4)/(6\sqrt{3})$$
$$P_{ij} = 1/12$$
$$\tag{3.9c}$$

Figure 3.11 shows a quantile plot comparison of the exact and approximate distributions of a quotient of two independent Normal variates, $y = x_1/x_2$, to illustrate equations 3.7, 3.8, 3.9. The variables have means and variances $\bar{x}_1 = \bar{x}_2 = 1$, $V(x_1) = 0.25$, $V(x_2) = 0.04$. The numerical integration method gives excellent agreement over the range of $P(y)$ from 0.001 to 0.999, the

Taylor's Series method gives good agreement over the range 0.01 to 0.90, while the Log-normal method is poor over most of the range.

Random sampling or simulation.   It is always possible to obtain the cdf of $y = f(x_i)$ empirically by substituting random sets of values of x_i, generated from the pdf's $g_i(x_i)$, into the equation and then determining $P(y)$ from the frequency distribution of y. The accuracy of this simulated distribution increases with the square root of the sample size, and consequently the sample will have to be quite large for useful results. Almost inevitably a digital computer must be available to generate the large sets of random variables and to calculate y for each set. Various methods can be used to economise the calculations (Hammersley and Handscomb (1964)), but this is purely a computational detail and will not be discussed further.

For a single variable x, the set of random variates can be calculated using random numbers z from a Uniform distribution, $0 \leqslant z \leqslant 1$, and the inverse of $P(x) = z$. This may involve interpolation in tables of $P(x)$ or a numerical approximation for the inverse.

For several dependent variables the generation of the sets is more difficult. For a bi-variate distribution of x_1, x_2 with joint pdf $h(x_1, x_2)$ use can be made of the conditional pdf

$$h(x_1, x_2) = g_1(x_1).g(x_2|x_1)$$

and a sample of x_1 taken which is then used to select a particular conditional pdf $g(x_2|x_1)$ for x_2 (Tocher (1963)). For example, with the Bi-Normal distribution of parameters $(\bar{x}_1, \bar{x}_2, \sigma_1^2, \sigma_2^2; \rho)$, taking two random standardised Normal variates u_1, u_2, with zero mean and unit variance, pairs of Bi-Normal variates would be given by

$$x_1 = \bar{x}_1 + u_1.\sigma_1$$
$$x_2 = \bar{x}_2 + \{\rho.u_1 + (1-\rho^2)^{\frac{1}{2}}.u_2\}.\sigma_2$$

Simulation of a distribution is usually a desperation measure, only to be resorted to if all other methods fail. Many of the published solutions of engineering problems treated by simulation could have been solved quite easily and cheaply by the methods discussed in the preceding sections.

Extreme values.   It is sometimes necessary to estimate the two extreme order statistics $x_{(1)}$ and $x_{(n)}$, the smallest and largest values in a sample of size n, from a population of cdf $P(x)$. The cdf's of $x_{(1)}$ and $x_{(n)}$ are obviously given by

$$P\{x_{(1)}\} = 1 - [1 - P(x)]^n$$
$$P\{x_{(n)}\} = [P(x)]^n$$

and these cdf's can be calculated for a known $P(x)$, although the final results will be very complicated. If $P(x)$ is bounded, so that $a \leqslant x \leqslant b$, then as n becomes very large the expected extremes approach the two bounds $x_{(1)} \to a, x_{(n)} \to b$, but for an unbounded distribution the extremes become very small or very large with increasing n, $x_{(1)} \to -\infty$, $x_{(n)} \to +\infty$.

Assuming that $P(x)$ is Normal, the cdf's for large n are asymptotically of the double exponential form

$$P\{x_{(1)}\} = 1 - \exp\{-\exp[(x-a)/b]\} \tag{3.10a}$$
$$P\{x_{(n)}\} = \exp\{-\exp[(x-a)/b]\} \tag{3.10b}$$

The most usual application of extreme values is for the case where the sample size n refers to time, and $x_{(n)}$ is the largest value expected to occur in a time n years, say. If the sample data of extreme values $y_1, y_2, \ldots, y_i, \ldots, y_n$, are the largest observed values y_i in i years, the cdf of y can be estimated, using equation 3.10b, by plotting y_i against $\log\{\log[1/P(i)]\}$, taking $P(i) = i/(n+1)$, and fitting a straight line to the data (Gumbel (1958)). Confidence limits can also be calculated for the line and hence for $P(y)$.

The recurrence period $T_i = 1/\{1 - P(i)\}$, the time during which an extreme y_i can be expected, is commonly used in practice instead of $P(i)$, but it must always be remembered that the value y_i could actually occur at *any* time, less than or greater than T_i, not necessarily at the end of the recurrence period itself.

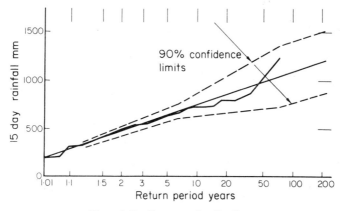

Figure 3.12   Extreme value distribution

Figure 3.12 gives an example of the confidence limits, for the total maximum rainfall occurring in Hong Kong over a period of 15 days (a factor influencing the frequency of slope failures), based on annual extremes measured over 74 years. With a confidence level of 90% the 100 year expected extreme would be in the range 750–1400 mm.

3.4 VARIABILITY OF SOIL PROPERTIES

For any extensive volume of natural soil or of fill, the processes governing the formation (sources of material, transporting and depositing agencies, etc.) will inevitably have fluctuated both temporally and spatially. For example, an estuarine deposit will have been controlled by material sources, currents, suspension and bed loads, etc., which may have had persistent components (e.g. river currents) modified by transient perturbations (e.g. tidal currents, wind effects) imposing a random pattern of variation onto an overall trend. The resulting soil properties can be regarded as being controlled by a random process. The variability may be small or large, depending on the relative importance of the persistent and transient effects, but in general there will be a greater tendency for the properties to be similar in value at closely neighbouring points than at widely spaced points.

When a property $\xi(x)$ varies continuously in value with distance x, then the mean and variance will depend on the range of x and the random process is *non-stationary*. However, in most practical cases it will usually be possible to convert ξ into a stationary process $\eta(x)$, whose mean and variance are sensibly independent of distance x, by taking d successive differences of ξ

$$\eta(x) = \nabla^d\{\xi(x)\}$$

where $\nabla(\xi) = \xi(x) - \xi(x - \delta x)$, is the Difference Operator (For an alternative treatment of non-stationary processes, see Matheron (1965)).

If ξ is measured at a series of n equally spaced distances $x_1, x_2, \ldots, x_i, \ldots, x_n$, so that $x_i = x_0 + i \cdot \delta x$, a general linear stationary random process can be written (Box and Jenkins (1970)) as

$$\varphi(T)\nabla^d\xi_i = \psi(T)\varepsilon_i + \mu$$

or
$$\varphi(T)\eta_i = \psi(T)\varepsilon_i + \mu \tag{3.11}$$

where ε is a random variable of zero mean and variance $V(\varepsilon)$;

        μ is a constant;

        T is the Translation Operator $T(\eta_i) = \eta_{i+1}$;

and φ and ψ are linear functions of T.

$$\varphi = 1 - \Sigma a_k T^k \quad ; \quad \psi = 1 + \Sigma b_k T^k \tag{3.12}$$

If b_k is zero the process is an *autoregressive process*, and if a_k is zero a *moving average random process*. Only autoregressive processes will be considered here.

The simplest process is the purely random uncorrelated process with $a_k = 0$

$$\eta_i = \mu + \varepsilon_i \tag{3.13}$$

and the next simplest, the unilateral Markov Process

$$\eta_i = a \cdot \eta_{i-1} + \varepsilon_i \tag{3.14a}$$

or
$$\eta_i = b \cdot \eta_{i+1} + \varepsilon_i \tag{3.14b}$$

where dependence extends only in one direction, either backwards (equation 3.14a) or forwards (equation 3.14b). For stability of the process the constants a or b must be less than unity.

The bilateral Markov Process with dependence extending in both directions is

$$\eta_i = a \cdot \eta_{i-1} + b \cdot \eta_{i+1} + \varepsilon_i \tag{3.14c}$$

In general, soil properties would be expected to show bilateral dependence when the direction x is the lateral direction, but certain properties would be expected to show only unilateral dependence on depth. Soil density and strength are strongly dependent on the effective overburden pressure which itself depends essentially on all the soil above the particular depth being considered. Consequently for such properties the depth dimension is in a sense a 'time-like' dimension with present values dependent on past values (i.e. at shallower depths) but not dependent on future values (i.e. at greater depths).

Since the property ξ will vary three-dimensionally with x, y, z (x and y horizontal, z vertical) the random process describing ξ will be a three-dimensional extension of equation 3.11.

$$\varphi(T_i T_j T_k)\eta_{ijk} = \mu + \varepsilon_{ijk} \qquad (3.15)$$

with η_{ijk} the value at x_i, y_j, z_k and

$$T_i(\eta_{ijk}) = \eta_{(i+1)jk}$$

$$T_j(\eta_{ijk}) = \eta_{i(j+1)k}$$

$$T_k(\eta_{ijk}) = \eta_{ij(k+1)}$$

The values of the coefficients in $\varphi(T_i T_j T_k)$ are related to the *autocorrelation coefficients* $\rho(stu)$ defined by equations such as

$$\sigma^2 . \rho(stu) = \text{cov}(stu) \qquad (3.16)$$

where
$$\sigma^2 = E(\eta_{ijk} - \mu)^2$$
is the variance and

$$\text{cov}(stu) = E\{(\eta_{ijk} - \mu)(\eta_{(i+s)(j+t)(k+u)} - \mu)\}$$

is the autocovariance for lags $s . \delta x$, $t . \delta y$, $u . \delta z$. For zero lag $\rho(000) = 1$.

For a representation such as equation 3.15 to be of practical use the number of terms in the series for $\varphi(T_i T_j T_k)$ should not be too large, which implies that $\rho(stu)$ decreases with increasing lag. Also, $\rho(stu)$ must be assumed to be independent of actual location x_i, y_j, z_k, and a function only of the lags.

The autocorrelation function is unlikely to be an isotropic function of R, the mean distance $R = (x^2 + y^2 + z^2)^{\frac{1}{3}}$. For very small lags (measured in millimetres) and very large lags (measured in kilometres) there is abundant evidence from petrographic or regional studies that $\rho(stu)$ is not even a circular-symmetric function of v, the mean horizontal distance $v = (x^2 + y^2)^{\frac{1}{3}}$. However, for lags of the order of a few metres, as in civil engineering projects, there is generally no strong evidence of preferred directions in the horizontal plane, and it may be assumed that $\rho(stu) = f_1(s) . f_2(t)$ where s now refers to a lag in any horizontal direction and t to a lag in the vertical direction. Thus ρ is assumed to be circular-symmetrical but not isotropic.

3.4.1 Autocorrelation functions

For the Random Uncorrelated Process analogous to equation 3.13 the autocorrelations are all zero for all lags except for zero lag

$$\rho(s) = 0 \qquad s > 0$$
$$\rho(s) = 1 \qquad s = 0$$

For the unilateral Markov Process of equations 3.14, it is easy to show that

$$\rho(s) = (\rho_0)^s$$

or, equivalently

$$\rho(s) = \exp(-\alpha s) \qquad (3.17)$$

For the bilateral Markov Process of equation 3.14c it can be shown (Whittle (1954), (1962)) that

$$\rho(s) = (\beta s) . K_1(\beta s) \qquad (3.18)$$

where $K_1(x)$ is the modified Bessel Function of the second kind and first order.

Thus the simplest form for a three-dimensional Markovian correlation function would be

$$\rho(s) = \exp(-\alpha t) . (\beta s) . K_1(\beta s) \qquad (3.19)$$

Sample autocorrelation coefficients $r(stu)$ are calculated from the results of tests at $N = n_1 . n_2 . n_3$ grid points located at

$$(x_1, y_1, z_1), \ldots, (x_i, y_j, z_k), \ldots, (x_{n_1}, y_{n_2}, z_{n_3})$$

preferably at equal spacings (for use of unequal spacings, see Agterberg (1970)), using equations of the form

$$\text{cov}(s) = \frac{1}{n_1 - s} \sum_{i=1}^{n_1 - s} \{\xi_{(i)} - \mu_{(i)}\} \{\xi_{(i+s)} - \mu_{(i+s)}\}$$

$$\sigma_{(i)}^2 = \frac{1}{n_1 - s} \sum_{i=1}^{n_1 - s} \{\xi_{(i)} - \mu_{(i)}\}^2$$

$$\sigma_{(i+s)}^2 = \frac{1}{n_1 - s} \sum_{i=1}^{n_1 - s} \{\xi_{(i+s)} - \mu_{(i+s)}\}^2$$

with
$$\mu_{(i)} = \frac{1}{n_1 - s} \sum_{i}^{n_1 - s} \xi_{(i)} \quad ; \quad \mu_{(i+s)} = \frac{1}{n_1 - s} \sum_{i}^{n_1 - s} \xi_{(i+s)}$$

and, finally,

$$r(s) = \frac{\text{cov}(s)}{\sigma_{(i)} . \sigma_{(i+s)}} \qquad (3.20)$$

and these sample autocorrelations averaged over all possible combinations of i, j, k, for all lags s, t, u.

Since sample autocorrelations are notoriously very sensitive to sampling variations a value of n_1, n_2, or n_3 of the order of 20 to 50 is the minimum to give a reliable estimate of $r(s)$. This implies a total number of tests N of the order of 10^4 or 10^5, which is prohibitively large even for a special research investigation.

A more realistic approach is to sample the soil properties along a few line transects in the x, y, z, directions, treating each set of results as a one-dimensional random process to estimate $\rho(s)$, $\rho(t)$, $\rho(u)$, and then to assume circular-symmetry if $\rho(s)$ and $\rho(t)$ are not too dissimilar. Even with this radical simplification the amount of testing is large and the arithmetical calculations require a digital computer, and consequently it is not surprising that very little has been published on autocorrelations.

The variance $V\{r(s)\}$ of the sample autocorrelations depends on the length of run of results, n, and on the population autocorrelations $\rho(s)$. If, for a lag s greater than q, say, the population autocorrelations are zero, the variance of $r(s)$ is approximately

$$V\{r(s)\} = \sigma^2 = \frac{1}{n}\left\{1+2\sum_{i=1}^{q}\rho_i^2\right\} \quad ; \quad s > q \qquad (3.21)$$

and $r(s)$ is approximately Normally distributed. Thus if $r(s)$ lies between the limits $\pm 2\sigma$ then ρ_s can be taken as zero, with a confidence level of about 95%.

Lateral dependence. A very extensive set of results for a clay shale, the Bear Paw Shale, were taken by the Prairie Farm Rehabilitation Administration, Saskatchewan, at spacings of 150 to 600 mm over a length of about 100 m along the walls of a tunnel (Bjerrum (1967)). *Figure 3.13* shows the sample autocorrelations for ξ and the first difference $\nabla\xi$ for liquid limits and liquidity index.

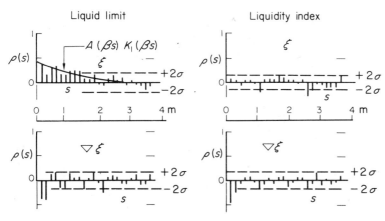

Figure 3.13   Sample autocorrelations for Bear Paw Shale

With a spacing of 150 mm, for the liquid limit results the first nine autocorrelations are non-zero but the values fall off with increasing distance more rapidly than would be predicted from equation 3.17 or 3.18. A combination of a Random Uncorrelated Process with a bilateral Markov Process is quite satisfactory, with

$$\rho(s) = 0.42(1.44s)K_1(1.44s) \quad ; \quad s > 0 \quad \text{measured in metres}$$

The first difference $\nabla\xi$ of the liquid limit gives a second order Autogressive Process $(1+0.62T+0.64T^2)\nabla\xi_i = \varepsilon_i$, with $V(\varepsilon) = (12.4)^2$. *Figure 3.14* shows a comparison between the values predicted by this equation, assuming ε Normally distributed, and a part of the actual observed series of results and agreement is quite good.

For the liquidity index results a Random Uncorrelated Process is satisfactory, since all autocorrelation coefficients can be taken as zero, except the 7th, 17th, and 18th (with 150 mm spacings). These small but non-zero values at large spacings are probably due to a slight cyclic trend.

A second example of lateral dependence is shown in *Figure 3.15*. These results refer to the modulus of subgrade reaction of a fine dry sand fill (quoted by Bolotin (1969, p. 152)) measured along a line transect by 162 loading tests

72

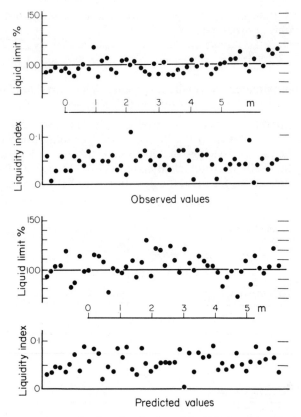

Figure 3.14   Observed and predicted values of liquid limit and liquidity index

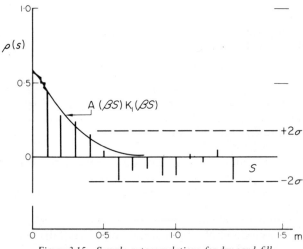

Figure 3.15   Sample autocorrelations for dry sand fill

on a 50 mm diameter plate at 100 mm spacings. Again, a combination of a Random Uncorrelated Process with a bilateral Markov Process gives a satisfactory representation.

$$\rho(s) = 0.59(6.35s)K_1(6.35s) \qquad ; \qquad s > 0 \quad \text{measured in metres}$$

Depth dependence. *Figure 3.16* shows the values of liquid limit, natural water content, effective density, and liquidity index of a normally con- solidated marine clay from Hong Kong (Lumb and Holt (1968)), measured on specimens taken at 300 mm spacings from a single borehole. For liquid

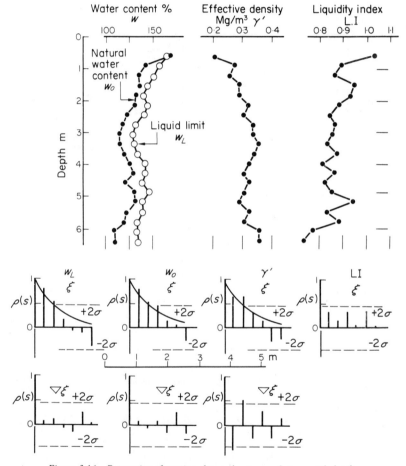

Figure 3.16   Properties of marine clay and autocorrelations with depth

limit, natural water content, and effective density a Unilateral Markov Process is satisfactory, with

$$\rho(t) = \exp(-1.23t) \qquad ; \qquad t \geqslant 0 \quad \text{in metres}$$

while for the liquidity index a Random Uncorrelated Process is good enough.

Figure 3.17 shows the undrained shear strength of this marine clay, measured by unconfined compression tests on specimens 50 mm dia. by

150 mm long taken at 300 mm spacings from several boreholes. Here, the autocorrelations decrease at a much slower rate with distance than those of any of the previous series, with

$$\rho(t) = \exp(-0.33t) \quad ; \quad t \geqslant 0 \quad \text{in metres}$$

A first order Unilateral Autogressive Process (UAP) represents the first difference $\nabla \xi$ quite adequately, with

$$(1 + 0.225T)\nabla \xi_i = \varepsilon_i \quad \text{kN}/m^2 \quad ; \quad V(\varepsilon) = (1.34)^2$$

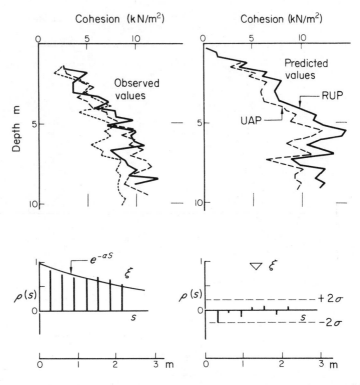

Figure 3.17   Observed and predicted cohesion of marine clay and autocorrelations with depth

although a simple Random Uncorrelated Process (RUP) is almost as good (*see Figure 3.17*), with

$$\nabla \xi_i = \varepsilon_i + 0.32 \quad ; \quad V(\varepsilon) = (1.33)^2$$

The two natural soils considered above, the Bear Paw Shale and the Hong Kong Marine Clay, are very 'uniform' deposits for which the spatial autocorrelations would be expected to be quite large. On the limited results given here the spatial autocorrelations are not particularly large even for these uniform soils, except for the depth dependence of strength, and no great error would be made in assuming the properties to be uncorrelated for spacings of the order of 1–2 m. Even for the strength of the marine clay, a Random Uncorrelated Process for the first difference is quite adequate and this implies that a linear dependence on distance of the form

$$\xi = \mu + ax + by + cz + \varepsilon$$

ought to be quite sufficient for most practical purposes.

While other soil properties for other soils may possibly be more strongly autocorrelated it can tentatively be assumed that a soil property may be regarded as the sum of a *trend component* $f(x, y, z)$ together with a purely *random component* ε of zero mean and variance $V(\varepsilon) = f_2(x, y, z)$, until further evidence becomes available. Thus estimation of soil variability reduces to estimation of the trend component and of the variance and probability density function of the random component.

3.4.2 Trend component

In routine site investigations there will rarely be sufficient test results available to estimate any form more complicated than a linear lateral trend, although the trend with depth may be investigated in more detail. A polynomial trend component

$$\xi = f(x, y, z) = \beta_0 + \beta_1 x + \beta_2 y + \beta_3 z + \beta_4 z^2 + \beta_5 z^3 + \text{etc} + \varepsilon$$

can be estimated using the method of Least Squares from the test results ξ_i at n locations (x_i, y_i, z_i). Rewriting $f(x, y, z)$ as

$$\xi = \sum_{j=0}^{k} \beta_j . x_j + \varepsilon$$

where x_j are the $(k+1)$ 'independent' variables $x_0 = 1$, $x_1 = x$, $x_2 = y$, $x_3 = z$, $x_4 = z^2$, $x_5 = z^3$, etc., or in Matrix notation

$$\xi = \mathbf{X} . \boldsymbol{\beta} + \varepsilon \tag{3.22}$$

then the estimates b_j of the true values β_j are given by

$$\mathbf{b} = (\mathbf{X}^\mathrm{T} . \mathbf{X})^{-1} . \mathbf{X}^\mathrm{T} . \xi \tag{3.22a}$$

where $\mathbf{X}^\mathrm{T}$ is the transpose of $\mathbf{X}$.

The estimate of $V(\varepsilon)$ is

$$\{n - (k+1)\} . V(\varepsilon) = \xi^T . \xi - \mathbf{b}^\mathrm{T} . \mathbf{X}^\mathrm{T} . \xi \tag{3.22b}$$

and the dispersion matrix (variances and covariances) of $\mathbf{b}$ is

$$V(\mathbf{b}) = V(\varepsilon) . (\mathbf{X}^\mathrm{T} . \mathbf{X})^{-1} \tag{3.22c}$$

If ε is assumed to be Normally distributed a $(1 - \gamma)$ confidence region for the whole trend component is

$$\mathrm{Prob}\left\{ \left| \frac{(\xi - \mathbf{X} . \mathbf{b})^2}{V(\varepsilon)\{\mathbf{X}^T (\mathbf{X}^T . \mathbf{X})^{-1} . \mathbf{X}\}} \right| \leqslant (k+1) . F^\gamma_{(k+1), n-(k+1)} \right\}$$
$$= 1 - \gamma \tag{3.22d}$$

where $F^\gamma_{v_1, v_2}$ is the upper tail value of Fisher's variance-ratio distribution for v_1 and v_2 degrees of freedom, such that

$$\mathrm{Prob}(F \geqslant F^\gamma_{v_1, v_2}) = \gamma$$

The deceptively simple form of equation 3.22 involves the inversion of a $(k+1)$ by $(k+1)$ matrix $(\mathbf{X}^\mathrm{T}.\mathbf{X})^{-1}$ and, if k is large, the arithmetic is very heavy.

For the simple relation $\xi = \beta_0 + \beta_1 x + \varepsilon$ the equations reduce to the well-known expressions

$$b_0 = \bar{\xi} - b_1.\bar{x}$$
$$b_1 = \Sigma(\xi_i - \bar{\xi})(x_i - \bar{x})/\Sigma(x_i - \bar{x})^2$$
$$V(b_0) = s^2/\Sigma(x_i - \bar{x})^2$$
$$V(b_1) = s^2\left[\frac{1}{n} + \frac{(\bar{x})^2}{\Sigma(x_i - \bar{x})^2}\right]$$
$$(n-2).s^2 = \Sigma(\xi_i - \bar{\xi})^2 - b_1^2.\Sigma(x_i - \bar{x})^2$$

The confidence region is bounded by a pair of hyperbolas above and below the regression line $b_0 + b_1.x$.

$$\xi_i = (b_0 + b_1.x_i) \pm s\left\{\frac{1}{n} + \frac{(x_i - \bar{x})^2}{\Sigma(x_i - \bar{x})^2}\right\}^{\frac{1}{2}} \cdot \{2F_{2,(n-2)}^\gamma\}^{\frac{1}{2}}$$

If only a limited range of x, $|x-\bar{x}| \leqslant a$, is of practical interest then a more useful confidence region is given by a pair of straight lines such that

$$\mathrm{Prob}\{|\xi - (b_0 + b_1.x)| \leqslant d.s\} = 1 - \gamma \tag{3.23}$$

although special tables (Bowden and Graybill (1966)) are necessary to find d.

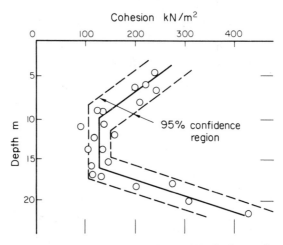

Figure 3.18   Confidence region for strength of cohesive till

Even when the trend is quite definitely non-linear it will often be satisfactory to assume piece-wise linearity over limited regions, for example by taking the variation with depth to be represented by a series of straight line segments, in order to simplify the calculations. *Figure 3.18* gives an illustration for the variation of undrained strength of a cohesive till with depth. The 95% confidence regions for the strengths were determined from equation 3.23 for each of the three line segments separately.

In the derivation of equation 3.22 the variance $V(\varepsilon)$ of the random component was assumed constant, independent of $\mathbf{X}$. If this variance is some known function, $V(\varepsilon) = \mathbf{V}.\sigma_0^2$ say, then equation 3.22 must be modified to

$$\mathbf{b} = (\mathbf{X}^T\mathbf{V}^{-1}\mathbf{X})^{-1}.(\mathbf{X}^T\mathbf{V}\xi)$$
$$V(\mathbf{b}) = \sigma_0^2.(\mathbf{X}^T\mathbf{V}^{-1}\mathbf{X})^{-1}$$

etc.

to give *weighted* estimates of parameters and dispersion matrix.

For the linear equation $\xi = \beta_0 + \beta_1 x + \varepsilon$ with a variance $V(\varepsilon_i) = (\sigma_0)^2/w_i$

$$b_0 = \xi_w - b_1.x_w$$
$$b_1 = \{\Sigma w_i(\xi_i - \xi_w)(x_i - x_w)\}/\Sigma w_i(x_i - x_w)^2$$

where ξ_w and x_w are weighted means

$$\xi_w = (\Sigma\xi_i.w_i)/\Sigma w_i \quad , \quad x_w = (\Sigma x_i.w_i)/\Sigma w_i$$

By calculating the 'residuals' $\varepsilon = \xi - \mathbf{X}.\mathbf{b}$ assuming $V(\varepsilon)$ constant and plotting these against the trend component $\mathbf{X}.\mathbf{b}$ and the independent variables $\mathbf{X}$, it will be possible to estimate the form of $\mathbf{V}$ in $V(\varepsilon) = \mathbf{V}.\sigma_0^2$ and to judge

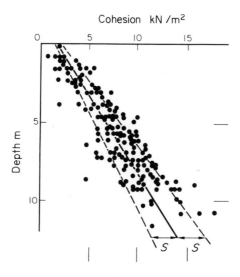

Cohesion kN /m²

Depth m

Figure 3.19 Strength of marine clay against depth

whether or not a weighted regression is necessary. In the case where only a linear trend with depth is assumed it will rarely be necessary to consider anything more complicated than a linear dependence of the standard deviation of the random component

$$\sigma(\varepsilon) = (\alpha_0 + \alpha_1.x).\sigma_0$$

or
$$w_i = (\alpha_0 + \alpha_1.x_i)^{-2} \tag{3.24}$$

with α_0 and α_1 estimated from a plot of ξ against x.

Figure 3.19 shows a more extensive set of results for the strength of the Hong Kong marine clay obtained from specimens from many boreholes.

Using equation 3.24, the trend component and the standard deviation of the random component are estimated to be

$$\text{Cohesion } \xi = 1.01(x+1.8)+\varepsilon \qquad \text{kN/m}^2, \quad x \text{ in metres}$$
$$\sigma(\varepsilon) = 0.19(x+1.8) \qquad \text{kN/m}^2$$

A linear variation for both trend and standard deviation as in *Figure 3.19* implies that the coefficient of variation $CV(\xi) = \sigma/\bar{\xi}$ is roughly constant. This constancy of coefficient of variation holds good for many soil properties,

Table 3.5 COEFFICIENTS OF VARIATION

Property	*CV%*
Density	5–10
Voids ratio	15–30
Permeability	200–300
Compressibility	25–30
Undrained cohesion (clays)	20–50
Tangent of angle of shearing resistance (sands)	5–15
Coefficient of consolidation	25–50

particularly for undrained cohesion, compressibility, and consolidation coefficient of clays, and some representative values (Lumb (1966), Shultze (1971)) are given in *Table 3.5*.

3.4.3 Random component

After determining the trend component, the pdf of the random component can be estimated provided that a sufficiently large sample of results is available. For each test result ξ_i, standardised random variables u_i can be calculated from

$$\mathbf{u} = (\xi - \mathbf{X} \cdot \mathbf{b})/\{V(\varepsilon)\}^{\frac{1}{2}}$$

and the pdf of $\mathbf{u}$ compared with an assumed population pdf using the methods of Section 3.3.

Most soil properties show a significant deviation from the Normal distribution at the extremes and are skew, but unless the sample size is large (of the order of 100) it will generally be difficult to choose a suitable alternative to the Normal. *Figure 3.20* shows some measured values of skewness and kurtosis to indicate the amount of deviation from the Normal distribution values $\sqrt{\beta_1} = 0$, $\beta_2 = 3$, that can be expected. The precision of these sample shape parameters is indicated by the size of the dispersion ellipses shown on *Figure 3.20*. For two variables X and Y of variances $V(X)$ and $V(Y)$, the equation of the dispersion ellipse is

$$\frac{x^2}{V(X)} + \frac{y^2}{V(Y)} = 1$$

Figure 3.20 shows that a bounded Pearson Main Type 1 distribution would give a closer fit to the residuals than would a Normal distribution. This is reasonable intuitively, since soil properties do have finite upper and

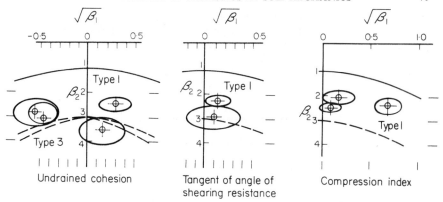

Figure 3.20   Shape parameters for soil property distributions

lower bounds to their possible range of values, and the infinite tails of the
Normal distribution must be rejected on physical grounds. On the other
hand, the Normal distribution usually gives a satisfactory fit to the central
range $(0.1 < P < 0.9)$ even for moderately skew distributions.

3.5   SAMPLING THEORY

The purpose of a site investigation is to determine those properties of the
soil which have a significant effect on the design, and from a statistical point
of view this implies determining sufficiently reliable estimates of means,
variances, and probability density functions. The amount of sampling in
routine investigations will inevitably be limited and the samples will always
be *small* samples (in a statistical sense), consequently estimation of the pdf's
may not be practical. Unless there is prior knowledge to the contrary it is
usual to assume that the population being sampled is a Normal population,
in order to make confidence statements about the estimates.

The first essential step is to determine the population to be sampled, and
this is by no means a trivial step. Preliminary investigations are intended to
delineate the different soil types in the region under study, but differentiation
on the basis of visual inspection or field tests may not be sufficiently precise.
This aspect has been emphasised by Morse (1971), in connection with
sampling of till formations where grouping all apparently similar formations
into one population would give quite misleading design parameters. This
problem is not limited to tills alone but arises with all soils.

Once the soil has been divided rationally into sub-regions, each of which
can be treated as an independent population on its own, then standard
techniques can be used to determine how many samples are required from
each sub-region and how the sampling should be performed.

3.5.1   Sampling patterns

Suppose samples are being taken from a single borehole through the
thickness H of any sub-region and a total of N samples are to be obtained.
These can be located at depths picked at random; picked systematically at

equally spaced intervals; or picked in what is called a stratified random pattern by dividing H into m sub-strata each of thickness h_i (i, 1 to m) with n_i samples allocated at random within each sub-strata. *Figure 3.21* illustrates these three sampling patterns, Simple Random (r), Systematic (sy), and Stratified Random (sr), for $N = 20$ and $m = 4$, with equal sample allocations $n_i = 5$.

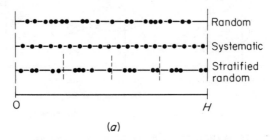

(a)

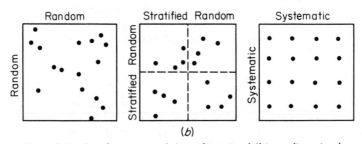

(b)

Figure 3.21   Sampling patterns (a) one-dimensional (b) two-dimensional

The relative efficiency of these three sampling patterns, as measured by the expected variance of the mean $E\{V(\bar{x})\}$, depends on the nature of the autocorrelation function $\rho(u)$. For a population variance $V(\xi)$, asymptotic values of the variance of the mean (Quenouille (1949)) are

$$V(\mathrm{r}) \qquad E\{V(\bar{x})\} = \frac{V(\xi)}{N}$$

$$V(\mathrm{sy}) \qquad E\{V(\bar{x})\} = \frac{V(\xi)}{N}\left[1 - \frac{2}{k}\int_0^{\infty} \rho(u)du + 2\sum_{u=1}^{\infty} \rho(k.u)\right]$$

$$V(\mathrm{sr}) \qquad E\{V(\bar{x})\} = \frac{V(\xi)}{N}\left[1 - \frac{2}{k^2}\int_0^{k} (k-u).\rho(u)du\right]$$

where k is the mean distance between samples.

For Markov Processes $\rho(u)$ is concave upwards, and it follows that

$$V(\mathrm{sy}) < V(\mathrm{sr}) < V(\mathrm{r})$$

and the systematic pattern is the most efficient of the three.

The same conclusions hold for sampling in two dimensions, where various combinations (r, r), (r, sy), (sy, sr), etc., of patterns could be used, as illustrated in *Figure 3.21(b)*. In general the systematic pattern in both directions will prove the most efficient.

If the soil property showed a cyclic trend component then systematic sampling would produce seriously biased estimates, since cyclic components of wavelength less than one half of the sample spacing would not be picked up at all, but this is unlikely to be a major source of error in routine investigations.

3.5.2 Sample size

The number of samples taken in each sub-region depends on the desired precision of the estimate. As given in Section 3.3, the mean and variance estimates from a sample of size n with test results x_i are

$$\bar{x} = \sum_i^n x_i/n$$

$$s^2 = \sum_i^n (x_i - \bar{x})^2/(n-1)$$

(3.25)

with the variance of the mean estimated by

$$V(\bar{x}) = s^2/n$$

(3.26)

Assuming Normality, confidence limits for the population mean ξ are given by

$$|\bar{\xi} - \bar{x}| = \frac{s}{\sqrt{n}} t^\gamma_{(n-1)}$$

where t^γ_v is 'Student's' t-variate with v degrees of freedom, such that

$$\text{Prob}\{|t| \geq t^\gamma_v\} = \gamma \qquad | \quad \gamma - two \ tailed$$

If the required precision Δ is $\Delta = (\bar{\xi} - \bar{x})/(\bar{x})$ and the sample coefficient of variation is $C = s/(\bar{x})$, then the minimum sample size to estimate the mean to within $\bar{x}(1 \pm \Delta)$ with confidence $(1 - \gamma)$ is

$$n = \left\{ \frac{C \cdot t^\gamma_{(n-1)}}{\Delta} \right\}^2$$

(3.27)

If no information on C is available before the investigation then a small sample of size n_0 should be taken as a preliminary to estimate C and then n found from equation 3.27. If n_0 turns out to be greater than n from equation 3.27 the sample is already large enough, otherwise a further $(n - n_0)$ samples are required.

The minimum sample size for $(1 - \gamma)$ 90, 95, and 99% is shown in *Table 3.6* as a function of Δ/C to indicate the relative importance of the various factors. If the coefficient of variation were about 20% then to estimate the mean to within $\pm 10\%$, with a confidence level of 95%, a sample size of 18 would be enough.

To estimate the variance with the same precision as the mean requires a much larger sample. Confidence limits for the population variance $V(\xi)$ are given by

$$(v \cdot s^2)/(\chi^2)^{\frac{1}{2}\gamma}_v < V(\xi) < (v \cdot s^2)/(\chi^2)^{1-\frac{1}{2}\gamma}_v$$

where $(\chi^2)_v$ is the Chi-squared variate with $v = (n-1)$ degrees of freedom. Usually only an upper confidence limit

$$V(\xi) < (v.s^2)/(\chi^2)_v^{1-\gamma}$$

is all that is needed and, to estimate the variance within

$$\{V(\xi)\}^{\frac{1}{2}} \leqslant s(1+\delta)$$

the minimum sample size is given in *Table 3.7*.

Table 3.6   MINIMUM SAMPLE SIZE FOR ESTIMATING MEAN VALUE

Δ/c		0.25	0.5	0.75	1.0	1.5	2.0
	99%	110	30	16	10	7	5
$(1-\gamma)$	95%	64	18	9	7	5	4
	90%	45	13	7	5	4	3

Table 3.7   MINIMUM SAMPLE SIZE FOR ESTIMATING STANDARD DEVIATION

δ		0.05	0.10	0.25	0.50
	99%	2000	600	65	24
$(1-\gamma)$	95%	750	160	36	14
	90%	400	100	24	10

3.5.3   Bayesian Estimation

In the previous sections it was assumed that nothing was known of the population prior to sampling, but in fact a considerable amount of information may be available in one form or another and it will be a very rare case where there is no prior information.

By using Bayes' Theorem of equation 3.3 the prior information can be incorporated into the analysis, giving more efficient estimates of the various parameters.

Suppose that the prior distribution of the population ξ is $N\{\bar{\xi}, V(\xi)\}$, meaning a Normal distribution with mean $\bar{\xi}$ and variance $V(\xi)$, and that n tests give estimates $\{\bar{x}, V(x)\}$. It can be shown that the posterior distribution of the *mean* $\bar{\xi}$ is $N\{\bar{\xi}^*, V(\bar{\xi}^*)\}$ where

$$\bar{\xi}^* = \frac{\bar{x}.V(\xi) + \frac{1}{n}.\bar{\xi}.V(x)}{V(\xi) + \frac{1}{n}.V(x)} \tag{3.28a}$$

$$V(\bar{\xi}^*) = \frac{1}{n}.V(x).\frac{V(\xi)}{V(\xi) + \frac{1}{n}.V(x)} \tag{3.28b}$$

As further information becomes available at later stages in the investigation, this can be incorporated through repeated use of equation 3.28, using the former posterior distributions as new prior distributions. When no prior information is available $V(\xi)$ can be taken as infinitely large, and equation 3.28 reduce to equation 3.25 and 3.26 as would be expected. Any prior information, however vague, will give $V(\bar{\xi}*)$ smaller than $V(\bar{\xi})$.

An extension of this approach (Tang (1971)) allows the inclusion of indirect measurements, in which ξ is estimated from measurements y through the use of a prediction equation, or calibration

$$E(\xi) = \beta_0 + \beta_1 . y, \quad \text{with variance} \quad V(\delta)$$

For a mixed set of n direct measurements and m indirect measurements, the posterior distribution of the mean $\bar{\xi}$ is $N\{\bar{\xi}**, V(\bar{\xi}**)\}$ with

$$\bar{\xi}** = \frac{\bar{x} . V(\mu) + \frac{1}{n} . \bar{\mu} . V(x)}{V(\mu) + \frac{1}{n} . V(x)} \tag{3.29a}$$

$$V(\bar{\xi}**) = \frac{V(x)}{n} . \frac{V(\mu)}{V(\mu) + \frac{1}{n} . V(x)} \tag{3.29b}$$

where, approximately†

$$\bar{\mu} = \frac{\frac{1}{m} \sum_{i}^{m} (\beta_0 + \beta_1 . y_i) . V(\xi) + \frac{1}{m} . \bar{\xi} . \{V(\xi) + V(\delta)\}}{V(\xi) + \frac{1}{m} . \{V(\xi) + V(\delta)\}} \tag{3.29c}$$

$$V(\mu) = \frac{1}{m} . \{V(\xi) + V(\delta)\} . \frac{V(\xi)}{V(\xi) + \frac{1}{m} . \{V(\xi) + V(\delta)\}} \tag{3.29d}$$

3.5.4 Censored and contaminated samples

With very soft or very stiff soils it may not be possible to obtain specimens at all the sample locations specified, or the specimens may be so badly disturbed that testing is useless. In such cases the set of data obtained will not be fully representative of the whole population and parameters estimated from the incomplete sample will be biased estimates of the population parameters.

If samples with properties less than ξ_0 or larger than ξ_1 cannot be taken at all, then the sampling is really from a *truncated* population of pdf

$$g(x) \qquad \xi_0 \leqslant x \leqslant \xi_1$$

$$g(x) = 0 \quad \left\{ \begin{matrix} x < \xi_0 \\ \xi_1 < x \end{matrix} \right\}$$

† The approximation involves neglecting the small effects of the variances of the parameters β_0 and β_1 in the prediction equation.

However, some information on the number of missing samples (with $x < \xi_0$, or $\xi_1 < x$) will be available from the boring records or laboratory reports, and the sample can be regarded as having been *censored* with m known values x out of a total of n, say, and with r_0 values less than ξ_0 and r_1 values larger than ξ_1, $m = n - (r_0 + r_1)$.

For singly censored samples (i.e. either r_0 or r_1 is zero) unbiased estimates of the population mean and variance take the form

$$\bar{\xi} = \bar{x} - \lambda(\bar{x} - \xi_0)$$
$$V(\xi) = s_0^2 + \lambda(\bar{x} - \xi_0)^2$$

where $\qquad \bar{x} = \sum_i^m x_i/m \qquad$ and $\qquad s_0^2 = \sum_i^m (x_i - \bar{x})^2/m$

and λ depends on the degree of censoring $h = r_0/(r_0 + m) = r_0/n$ and on $s_0^2/(\bar{x} - \xi_0)^2$.

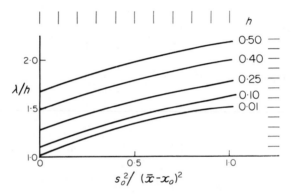

Figure 3.22 λ Factor for censored samples

For a Normal population, tables of λ are given by Cohen (1961) and some typical values are shown in *Figure 3.22*. If h and $s_0^2/(\bar{x} - \xi_0)^2$ are small then λ is close to h, and the censored values can approximately be replaced by r_0 values at the truncation point $x = \xi_0$.

If the censoring is severe (h large) then better estimates can usually be made through a prediction equation if the wanted property can be related to soil tests performed on disturbed specimens (i.e. liquid and plastic limits, natural water contents), using equation 3.29.

If the boundaries of the sub-regions are ill-defined or have been chosen carelessly there is a strong possibility that the sample will be a mixture of results relating to two or three populations, or the sample is *contaminated*. All suspect values should be eliminated from the sample before calculating estimates, otherwise interpretation through confidence limits, etc., will have very doubtful validity (see Morse (1971)). In large samples, contamination will produce a multi-modal distribution, and removal of the spurious data involves decomposition of the pdf as described in Section 3.3.

In small samples, where there is contamination by one additional population, there will only be two or three values which are spurious and these will generally be at the extreme of the range, either suspiciously small or suspiciously large in comparison with the rest of the data. The simplest method for dealing with such possible 'outliers' is to calculate the ratio

$$T_n = |x_{(n)} - \bar{x}|/s$$

where $x_{(n)}$ is the largest (or smallest) value and to remove $x_{(n)}$ if T_n is greater than a critical value, repeating the process for $(n-1), (n-2)\ldots(n-r)$, until T_{n-r} is less than the corresponding critical value. For a Normal population,

Table 3.8   CRITICAL VALUES OF T_n FOR OUTLIERS

	n	3	5	10	15	20	25
	99%	1.15	1.75	2.41	2.71	2.88	3.01
$(1-\gamma)$	97.5%	1.15	1.71	2.29	2.55	2.71	2.82
	95%	1.15	1.67	2.18	2.41	2.56	2.66

critical values T_n are given in *Table 3.8* (after Grubbs (1969)) for a one-sided test (i.e. the outliers are known to be on the high side or on the low side). For two-sided tests (i.e. the outlier could be either high or low) the significance levels of *Table 3.8* should be doubled.

3.6   PRECISION AND ACCURACY OF SOIL TESTS

It has been assumed so far that tests on soil specimens give exact values of the desired property but this is not necessarily true, and a test result will usually contain an 'error' component.

For a true value ξ and corresponding test value x, repeated or replicate testing on the *same* specimen may give different values of x because of lack of *precision* of the test, while the average of a number of replicate tests may differ from the true mean because of lack of *accuracy* of the test.

For the i-th replication the test value x_i can be written

$$x_i = \alpha.\xi + \beta + \delta_i$$

where δ is a random variable of zero mean and variance $V(\delta)$.

The average of a large number of tests, the expectation, is

$$E(x) = \alpha.\xi + \beta \qquad (3.30)$$

and α and β express the *bias* or lack of accuracy, while $V(\delta)$ represents the lack of precision, the larger the variance the lower the precision.

Quite clearly, there will be no way of determining the bias factors of equation 3.30 unless some absolute method of measuring ξ is available but for most soil tests there is no such absolute method, many properties being in fact defined as the result of a 'standard' test performed with specified equipment under specified conditions.

The true value ξ is occasionally inferred using calibration or model tests in which some response ζ is predicted by a theoretical equation

$$\zeta = f(\xi)$$

The measured response z is compared with the measured value x, but bias effects are inevitably co-mingled since both $f(\xi)$ and z may themselves be biased. In practice, the bias will be lumped either into the response model or into the test, giving

$$\text{Model Bias} \quad \begin{aligned} E(z) &= \alpha_m \cdot f(\xi) + \beta_m \\ E(x) &= \xi \end{aligned} \tag{3.31a}$$

$$\text{Test Bias} \quad \begin{aligned} E(z) &= f(\xi) \\ E(x) &= \alpha_t \cdot \xi + \beta_t \end{aligned} \tag{3.31b}$$

and perhaps the most satisfactory treatment of accuracy is to take one well-established standard test as an 'absolute' test, giving an unbiased estimator of ξ and to attribute all inaccuracies to the response equation as in equation 3.31a.

Changes from the specified test conditions (use of non-standard equipment, non-standard environment, or use of unskilled operators) will produce deviations in the results which can be interpreted as either bias factors or precision factors. For the same specimen n replicate tests are performed by p operators each using a set of q different test machines. The k-th repeat test by the j-th operator on the i-th machine can be written (for a test unbiased on average) as

$$x_{ijk} = \xi + \alpha_i + \beta_j + \gamma_{ij} + \delta_{ijk} \tag{3.32}$$

$$i, 1 \text{ to } q; \quad j, 1 \text{ to } p; \quad k, 1 \text{ to } n$$

where α_i represents the Machine Effect $E(\alpha_i) = 0$

            β_j represents the Operator Effect $E(\beta_j) = 0$

            γ_{ij} represents the Interaction between machine and operator $E(\gamma_{ij}) = 0$

and δ_{ijk} is a random variable of zero mean and variance $V(\delta)$

If there are a fixed number of operators or machines (as in any one particular testing laboratory) then α_i, β_j, γ_{ij}, can be regarded as fixed constants or bias terms associated with a particular machine and operator. If the machines and operators are picked as a random sample from the whole possible population of machines and operators (choosing a laboratory at random) then α, β, γ, can be regarded as random variables having variances $V(\alpha)$, $V(\beta)$, $V(\gamma)$, and the effect of these variables is to inflate the imprecision.

3.6.1   Test replication

Inter-laboratory tests, in which the same standard specimens are tested repeatedly by many operators in different laboratories, are necessary in order to separate the various effects, and estimation of the factors carried out through Analysis of Variance (Davies (1967), Mandel (1964)), but such comprehensive studies have yet to be done for soil testing and a simpler model than equation 3.32 will be used, lumping all three effects together so that the j-th repeat test in the i-th laboratory is

$$x_{ij} = \xi + \alpha_i + \delta_{ij} \quad ; \quad j, 1 \text{ to } n; \quad i, 1 \text{ to } k \tag{3.33a}$$

For one particular laboratory the mean value $\bar{x}_i$ is an estimate of

$$E(\bar{x}_i) = \xi + \alpha_i \quad ; \quad V(\bar{x}_i) = \frac{1}{n} \cdot V(\delta) \quad\quad (3.33\text{b})$$

while for the entire set of all laboratories the grand mean $\bar{x}$ is an estimate of

$$E(\bar{x}) = \xi \quad ; \quad V(\bar{x}) = \frac{1}{k} \cdot V(\alpha) + \frac{1}{nk} \cdot V(\delta) \quad\quad (3.33\text{c})$$

Obviously, if $V(\alpha)$ is much larger than $V(\delta)$ then test replication can have little influence on reducing the variance of the mean. (χ²-square— Analysis of Variance)

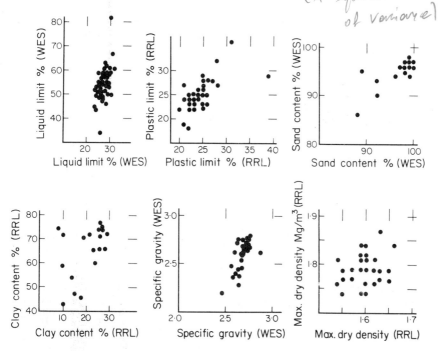

Figure 3.23   Inter-laboratory variability of classification tests

Classification tests. Two fairly extensive inter-laboratory studies have been carried out by the Waterways Experimental Station (WES) (Hammitt (1966)), and the Road Research Laboratory (RRL) (Sherwood (1970)). For each study three different soil types were sent as 'standard' specimens to each of 100 (WES) and 40 (RRL) commercial, research, or government laboratories. Atterberg limits, particle-size distribution, specific gravity, and standard and modified compaction tests were performed on each specimen by each laboratory.

The individual 'error' terms $(\alpha_i + \delta_{ij})$ were much larger than would be expected intuitively for such well-known routine tests, as is illustrated in *Figure 3.23* where pairs of results for tests on two of the soils are plotted for each laboratory. Analysis of these and other studies (Lumb (1971b)) shows that in general α and δ depend on the actual magnitude of ξ, but can be characterised roughly by coefficients of variation. *Table 3.9* gives weighted mean values of

$$CV(\alpha+\delta) = \{V(\alpha)+V(\delta)\}^{\frac{1}{2}}/\bar{\xi}$$
$$\text{and} \qquad CV(\delta) = \{V(\delta)\}^{\frac{1}{2}}/\bar{\xi}$$

corresponding to expected values for a competent commercial laboratory.

Test replication has little effect on reducing the error term; for example, using the mean of three repeat tests the coefficient of variation of the liquid limit would be reduced from 6.0% to 5.9%, and that of the plastic limit from 8.7% to 8.4%, both being negligible gains in precision.

The main source of the error is deviation from the specified test method, and if precise classification test results are needed then very strict test control is essential. If the test results are to be used in empirical design equations (e.g. pavement design by the Group Index Method) or for quality

Table 3.9 MEAN PRECISION COEFFICIENTS OF VARIATION FOR CLASSIFICATION TESTS

Test type	$CV(\alpha+\delta)\%$	$CV(\delta)\%$
Liquid limit	6.0	1.2
Plastic limit	8.7	2.9
Sand content	1.1	—
Clay content	11.4	0.7
Specific gravity	1.4	0.2
Maximum dry density (standard)	1.6	0.7
Maximum dry density (modified)	1.5	0.8

control (e.g. plasticity index for fill material) it is extremely important that the test specifications be rigidly adhered to, particularly if several test operators have to be employed to carry out the testing on one job.

Significance of test errors.   When classification or other tests are carried out to estimate the population values of a soil sub-region (as in Section 3.5) the test errors may have little significance, since the 'site variance' of a nominally uniform soil is much larger than the single-test variance $\{V(\alpha)+V(\delta)\}$. Suppose one operator performs n repeat tests on each of N specimens from a site where the soil property ξ is $\{\bar{\xi}, V(\xi)\}$, and the j-th repeat test on the i-th specimen gives a result x_{ij}

$$x_{ij} = \xi_i+\alpha_i+\delta_{ij}$$
$$\text{with} \qquad E(\bar{x}_i) = \xi_i+\alpha_i$$
$$E(\bar{x}) = \bar{\xi}+\bar{\alpha}$$

where $\bar{\alpha}$ is a mean bias for this operator.

The variance of the specimen means

$$V(\bar{x}_i) = \sum_{i}^{N} (\bar{x}_i-\bar{x})^2/(N-1)$$

has an expected value

$$E\{V(\bar{x}_i)\} = V(\xi)+V(\alpha)+\frac{1}{n}.V(\delta) \qquad (3.34a)$$

and, if the operator is consistent, then the bias α_i is sensibly constant and approximately $V(\alpha) = 0$, leading to

$$E\{V(\bar{x}_i)\} = V(\xi) + \frac{1}{n}.V(\delta) \qquad (3.34b)$$

analogous to equation 3.33c.

To illustrate the relative importance of site and test variances *Table 3.10* gives some average values of coefficient of variation for a 'site', from $\{V(\xi) + V(\delta)\}$, and a single test ($n = 1$), from $V(\delta)$ derived from undisturbed

Table 3.10   AVERAGE SITE AND TEST COEFFICIENTS OF VARIATION FOR UNDISTURBED SOIL

Property	Site CV%	Single test CV%
Permeability (sand)	240	22
Compression index (clay)	30	10
Consolidation coefficient (clay)	47	32
Undrained cohesion (clay)	27	14

Table 3.11   PERCENTAGE REDUCTION IN SITE COEFFICIENT OF VARIATION

Soil property	Number of repeat tests			
	2	3	5	∞
Permeability	0.2	0.3	0.3	0.4
Compression index	2.8	3.8	4.5	5.7
Consolidation coefficient	12.4	16.9	20.7	26.8
Undrained cohesion	7.0	9.4	11.4	14.5

specimens from typical sites (Lumb (1971b)). The average gain in precision obtained by replication is shown in *Table 3.11* as the percentage reduction in site coefficient of variation CV(site)

$$CV\text{(site)} = \left[V(\xi) + \frac{1}{n}.V(\delta) \right]^{\frac{1}{2}} / \bar{\xi}$$

The only significant gains are for consolidation coefficient and for undrained cohesion, for both of which three to five replications give as much reduction in CV as is economically warranted.

Soil strength.   Routine soil strength tests are performed by shear-box, drained triaxial, or consolidated-undrained triaxial tests and, for cohesive soils, by undrained triaxial (UU), unconfined compression (U), laboratory vane (LV), or field vane (FV) tests. Usually the amount of testing for one site is small enough for a single laboratory to cope with the work and consequently the α-component of equation 3.34 can be treated as a bias factor with the precision measured by the δ-component.

Table 3.12 gives average site and test coefficients of variation for undisturbed specimens of clay shale, cohesive till, and residual sands and silts, all soils which are notoriously extremely variable in strength. For these cohesive-frictional soils the cohesion is far more variable than the angle of shearing resistance, but three to five replications again give about as much reduction in test variance as is warranted. There is little information on the relative precision and accuracy of shear-box and triaxial tests, but on the whole the triaxial is rather more precise than the shear-box (Lumb (1971b)) while the accuracy is about the same.

For undrained strength of undisturbed clays there is more information available, and *Figures 3.24 and 3.25* show comparisons between FV and LV tests, and between U or UU and LV tests for soft to firm saturated clays. The 95% confidence regions of *Figure 3.24* include the equal-accuracy line

$$FV = LV$$

and it may be safely concluded that FV and LV tests give equivalent results. However, in *Figure 3.25* the equal-accuracy line

$$U, UU = LV$$

Table 3.12   COEFFICIENT OF VARIATION FOR STRENGTH

Soil type	Test	Property	Site CV%	Single test CV%
Clay shale	Shear box	cohesion	95	44
		$\tan \varphi$	46	22
Cohesive till	Shear box	cohesion	103	72
		$\tan \varphi$	18	6
Residual sands and silts	Triaxial	cohesion	17	13
		$\tan \varphi$	6	2

is excluded from the confidence regions over most of the range of strength and hence the U or UU tests are biased relative to the LV or FV tests. The mean relative accuracy, for strengths up to 80 kN/m² is

$$U, UU = 0.73(LV, FV)$$

The relative precision of LV, U, and UU tests is about the same, and the influence of test replication on variance reduction given in *Table 3.11* holds good, on average, over the full range in strength from very soft to very stiff.

The main source of the relative bias of UU and LV tests is, of course, disturbance of fabric or structure during sampling and handling. Extremely careful sampling can produce equivalent accuracies for FV and U test (c.f. Lumb and Holt (1968)) but the special equipment required is rarely used in routine investigations.

3.6.2   Prediction equations

The same concepts of precision and accuracy also apply to indirect estimation through the use of prediction or calibration equations relating the wanted property to some more easily measured property.

Estimation of compression index or consolidation coefficient from the liquid limit, of cohesive strength from the standard penetration test, and so on, can be treated on the lines of equation 3.31a, regarding the predictor

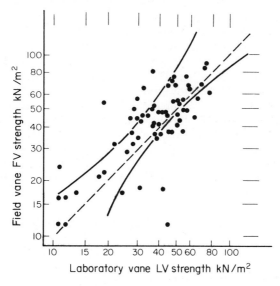

Figure 3.24 Comparison of field and laboratory vane strengths

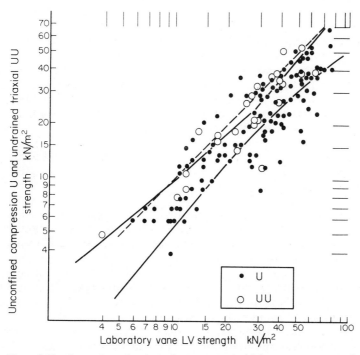

Figure 3.25 Comparison of undrained compression and laboratory vane strengths

variable (liquid limit, standard penetration N-value) as known exactly, but the calibration as being biased and imprecise. Provided that the bias is known, then imprecise predictors can be used to give quite precise estimates if sufficient replication tests are carried out (c.f. equation 3.33b). The difficulty lies in estimating the bias, which will almost always vary with soil type, etc., and cannot be predetermined for all site conditions.

By including more than one predictor variable in the equation the bias can often be stabilised and the precision increased, but at the cost of additional preliminary calibration and of a more complicated predictive equation (see Holtz and Krizek (1971) for details of multi-variate predictors). It is always important, of course, that these empirical predictor equations are not extrapolated too far beyond the validated calibration range.

3.7   SPECIFICATION AND CONTROL

For construction on undisturbed soil the essential design problem is estimating the properties and behaviour of the existing soil, but when designing earth dams, embankments, roads and pavements, the type of material and its properties can be specified in advance by the designer and the problem here is to ensure that the fill as placed will behave in the specified manner. There is some degree of control over the materials and the construction method and unsatisfactory fill can be modified or replaced if necessary.

There are two main parties involved in construction work, the client or designer, and the contractor or supplier, whose objectives are not quite the same. From the client's point of view there must be a minimum risk of accepted work being actually below specification, while from the contractor's point of view there must be a minimum risk of rejected work actually being up to specification. Compromise between the two points of view is essential and, for both parties to be satisfied, it is important that any inspection criteria for rejection or acceptance must be based on realistic specifications, paying due regard to unavoidable variations in materials, methods, and inspection. Statistical quality control schemes can be of great value in achieving this.

3.7.1   Earthwork control

The performance of compacted earthwork is predicted on the basis of past experience with similar soils (prior information) up-dated by laboratory testing and field trials on available materials. The stability and deformations of the completed work are functions of the variable strengths and compressibilities of the compacted fill, but these parameters are not currently specified for acceptance criteria. This is for the obvious reason that such tests require a relatively long time to perform, whereas a decision on acceptance is needed quickly to avoid excessive delay on construction.

For the materials, acceptance is usually based on limiting ranges of particle-size distribution or limiting ranges of the plasticity index. For state of compaction, acceptance is usually based on the dry density being greater than a certain fraction of maximum dry density, and on water content being within a limiting range wet- or dry-of-optimum water content. Since both maximum dry density and optimum water content depend on material type, control compaction tests on all accepted materials are strictly necessary in order to assess the state of compaction. An alternative specification basing acceptance on a maximum allowable percentage of air voids in the compacted fill partly eliminates the effects of material variability, and simplifies the inspection acceptance scheme.

The specified acceptance criteria ought to be set with due regard to the overall variability which occurs in satisfactory earthworks. From measurements on earth dams and road embankments (Davis (1953), Turnbull *et al.* (1966), Smith and Prysock (1966), Pettit (1967), Beaton (1967)) the overall variability of perfectly satisfactory fill is equivalent to a dry density coefficient of variation ranging from about 2% to 10%, depending essentially on material variability, and a placement water content standard deviation of 1 to 3%. In terms of percentage air voids, an acceptable overall variability (RRL (1967)) is equivalent to a standard deviation of 3 to 4.5%.

Table 3.13 ACCEPTANCE VALUES

	$\alpha\%$	20	10	5	1	0.5	0.1
Dry density	$\Delta/\xi^{do}\%$	4.2	6.4	8.2	11.6	12.9	15.5
Water content	$\Delta\%$	2.6	3.3	3.9	5.2	5.6	6.6
Air voids	$\Delta\%$	3.2	4.9	6.3	8.8	9.8	11.7

Suppose that the design parameter ξ_d is the *mean* value and that the variance of the compacted fill is $V(\xi) = \sigma^2$. If α is the probability of the material being below the acceptance value then

$$\xi_\alpha = \xi_d + \sigma . u$$

where u is the standardised variate for ξ. Assuming Normality, the acceptance values for dry density ($\xi_\alpha = \xi_d - \Delta$), placement water content ($\xi_\alpha = \xi_d \pm \Delta$), or percentage air voids ($\xi_\alpha = \xi_d + \Delta$) would be as shown in *Table 3.13*. For example, setting acceptance limits to be greater than 95% of the design value for dry density, water content to be within $\pm 3\%$ of design value, or percentage air voids to be less than design value plus 5%, would result in about 10% of perfectly acceptable fill being rejected if the contractor was actually achieving a mean value equal to the design value. To avoid this rejection the contractor would have to aim at a much tighter control on compaction than is really necessary to guarantee adequate performance. Thus, for a rejection percentage of 1% instead of 10% the contractor would have to aim at a mean dry density of about 105% of the design value.

Once realistic specifications have been made, the client's interests must be safeguarded by site control tests and inspection. There are two main ways in which control can be applied (Grant (1964)), control by attributes and control by variables.

Control by attributes. The whole unit of work (one layer of fill, one shift output, etc.) is accepted if at most n out of N tests on a unit fall outside the specification limit ξ_α. For an acceptance probability α the probability of obtaining *at most* n results out of N less than ξ_α is given by the Binomial distribution $P(n\,|\,N,\alpha)$

$$P(n\,|\,N,\alpha) = \sum_{r=0}^{n} \binom{N}{r} \cdot (\alpha)^r \cdot (1-\alpha)^{N-r}$$

and hence the probability β of rejecting the unit when it is in fact within specification is

$$\beta = 1 - P(n\,|\,N,\alpha)$$

For samples of size $N = 5$, 10, and 20, the values of n to give at most 5% for β are shown in *Table 3.14* for different acceptance probabilities α. Thus, if 2 out of 10 results fail to pass the specification, for example, this would be sufficient reason for rejecting the unit if the acceptance probability were 5% but insufficient reason for rejection if the acceptance probability were 10%.

Table 3.14   NUMBER OF REJECTS
FOR $\beta \leqslant 5\%$

Sample size	$\alpha\%$	5	10	20	30
5	—	1	2	3	3
10	—	2	3	4	5
20	—	3	4	7	8

With fixed values of α, β, and sample size N, the number of allowable rejects n is also fixed and hence rejection or acceptance of a unit is a simple decision matter.

Control by variables.   If the individual test results x_i in the sample of size N are used then more information on the population is available, and acceptance can be made contingent on the sample mean $\bar{x}$ and sample variance s^2 being within specified limits.

Referring to *Figure 3.26*, if ξ_L and ξ_U are lower and upper acceptance limits such that $\text{Prob}(\xi_L < \xi < \xi_U) = 1 - \alpha$, and A and B represent lower and upper acceptable populations, then for the sample mean the lower and upper control limits LCL and UCL cover a narrower range than ξ_L and ξ_U, since the variance of the mean is $V(\xi)/(N) = \sigma^2/N$

$$\text{LCL} = \xi_L + k_1 \cdot \sigma - k_1 \cdot \sigma/(N)^{\frac{1}{2}}$$
$$= \xi_L + k_1 \cdot \sigma(1 - 1/N^{\frac{1}{2}})$$
$$\text{UCL} = \xi_U - k_2 \cdot \sigma(1 - 1/N^{\frac{1}{2}})$$

with k_1 and k_2 fixed for a given probability α.

The population variance is estimated from the sample variance s^2, and k_1 and k_2 determined using 'Student's' t-variate (e.g. Kühn (1971)). For small samples (N of the order of 5) it will be commonly sufficiently precise to estimate the variance from the sample range $R = (x_{\text{max}} - x_{\text{min}})$.

Since control is being based on an acceptable mean value for the whole unit, the sampling should be truly representative of the unit, either systematic sampling at regular spacings or stratified random sampling being used.

Sampling at specific points selected after initial inspection is not recommended, because of the danger of producing a heavily biased sample.

The size of the sample should be related to the actual volume of the unit and, with constant thickness of uncompacted material, a size of the order of one test per 400 to 800 m^2 is usually adequate. If the process is under satisfactory control, occasional sample values outside the control limits are no cause for alarm but persistent variations and trends are a warning that the process is changing.

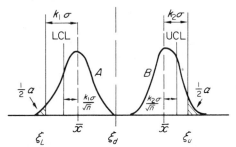

Figure 3.26   Control limits for the sample mean

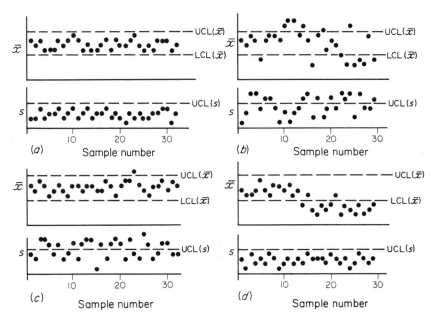

Figure 3.27   Control charts for mean and standard deviation

By plotting the sample values of $\bar{x}$ and s (or R) sequentially on a control chart marked with the control limits LCL and UCL, as in *Figure 3.27*, any significant changes in the process can be picked up quite easily. The set of results shown in *Figure 3.27a* is well in control, with $\bar{x}$ and s both within the control limits, while the set of *Figure 3.27b* is seriously out of control for both $\bar{x}$ and s. This lack of control could be due to material variability, lack of adequate compaction supervision, or unrealistic specifications resulting in unattainable control limits. In *Figure 3.27c* the mean is under control but s is not, most likely because of inconsistent compaction methods. In *Figure 3.27d* the standard deviation is under control but the mean shows a shift in level after the 15th sample, most likely due to a change in material.

3.8   STATISTICAL DESIGN

The analytical part of a design relating the response to the loads and soil properties is inevitably approximate, since many simplifications must be made in order to produce a solution. No matter what method of analysis is used, characteristic design values of loads and properties have to be allocated to the sub-regions, and in current practice it is assumed that these characteristic values are in fact known or knowable constants. Design 'errors' due to idealisation of soil behaviour (elastic, rigid-plastic, etc.) can in principle be assessed through model tests on uniform soils, but the influence of variabilities is not easy to determine.

If ζ^* is the analytical response assuming *constant* loads η_i and properties ξ_i for each sub-region, then with variable loads and properties having sub-regional means $\bar{\eta}_i, \bar{\xi}_i$, the actual response ζ will in general be

$$\zeta = \alpha . \zeta^* + \beta + \delta \qquad (3.35)$$

where $\zeta^* = f(\bar{\eta}_i, \bar{\xi}_i)$ and δ is a random variable $\{0, V(\delta)\}$, (c.f. equation 3.31). Using the methods of Sections 3.3, 3.4, 3.5 and 3.6, the pdf of ζ could be found and the effects of soil variability interpreted as bias factors α and β and precision factor δ. In general neither bias factors nor precision variance will be constants but will be functions of $\bar{\eta}, \bar{\xi}$.

A combination of the finite element method and the simulation method of Section 3.3 may give useful insight into the factors of equation 3.35. Kraft and Yeung (1973) estimated the mean and variance of the elastic deformation of a soil slope, taking the modulus of elasticity E as a random Normal variate and assuming all other parameters to be constants. *Figure 3.28* shows typical results for the vertical and horizontal deformations s_v and s_h in terms of the coefficient of variation $CV(E)$ of the modulus of elasticity. The mean response $\bar{\zeta} = \alpha . \zeta^* + \beta$ increases slightly with $CV(E)$ implying that the softer elements have a greater influence than the stronger elements, and both $\bar{\zeta}$ and $CV(\zeta)$ depend non-linearly on $CV(E)$.

3.8.1   Characteristic loads

Some loads can always be treated deterministically as known constants whereas others are truly random variables and ought to be treated probabilistically. The usual distinction between dead loads and live loads is in fact a consequence of this, and while the demarcation between dead and live loads may not always be clear-cut nevertheless the distinction is a useful one.

Dead loads.   Gravity loads due to self-weight are proportional to the mean soil density above a particular depth. The coefficient of variation of density is relatively small, of the order of 5 to 10% (c.f. *Table 3.5*), and the coefficient of variation of the *mean* density will almost always be negligibly small in comparison with other variables. No significant errors will be caused by ignoring this small variability and by treating gravity loads deterministically.

Structural dead loads due to buildings, fill, dams, etc., can again be treated deterministically provided the dimensions of the proposed structure are known in advance. If there is a strong possibility of the loads being increased eventually, by additional storeys to a building, heightening a dam, etc., this possibility ought to be taken into account by using the expected *maximum* dead load for a given working life.

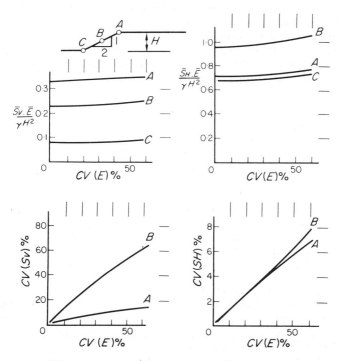

Figure 3.28   Elastic deformations of slope, mean values and coefficients of variation

Live loads. Earthquake loads must be treated as random variables in any rational design, and choice of a suitable characteristic load is a matter of estimating the expected *extreme* loads occurring during earthquakes of various magnitudes and also estimating the likelihood of such earthquakes. The characteristic load will consequently be associated with a return period, as described in Section 3.3. There will always be a finite risk of the actual load exceeding the characteristic load during the return period, and thus a very real risk of over-stressing during the expected working life.

For structural live loads (building floor loads, bridge traffic loads, etc.) the characteristic values used in designing the structure itself are commonly far larger than the loads actually encountered during the working life. When there is some control on the use of the structure (maximum allowable floor load, maximum allowable axle-load, etc.) then these control limits can be used deterministically to estimate extreme loads imposed by the structure on the soil, but for uncontrolled use the characteristic loads will be expected extremes associated with a return period as in the case of earthquakes.

Water loads.   Pore-water pressures can be treated deterministically in some instances but not in others. For example, with earth dams the extreme upper and lower expected pore-water pressures can often be estimated quite precisely, or the actual values measured during construction and these measured values incorporated into the design. In cuttings and natural slopes, however, precise estimation is almost impossible since the pore-water pressures depend on many uncontrollable factors such as sources of re-charge water, drainage patterns, and rainfall.

Broadly speaking, the case of a 'static ground-water table' is a rarity, and pore-water pressure is governed by a time-dependent random process. Characteristic values will again be associated with a return period and should be estimated from expected extreme values.

3.8.2   Characteristic soil properties

Although the major design problems in soil engineering are really problems in estimating deformations, it is customary to consider two separate classes of *settlement* (foundation settlements) and of *stability* (bearing capacity, slope stability). For the first class of problems the soil compressibility controls the design while for the second class the soil strength controls. In principle it would be possible to measure the soil properties everywhere in the affected volume, but in practice this is quite impossible. Estimates have to be made on the basis of small numbers of tests and as far as the designer is concerned the properties must be considered as random variables.

Soil strength.   The stress-strain curves of natural soils are typified by *Figure 3.29*, stress increasing with strain up to a peak value ξ and then decreasing to a final residual value $r \cdot \xi$ ($r \leqslant 1$). In any sub-region the peak strength will be a random variable of pdf $g(\xi)$, and the applied stresses assumed to be constant characteristic values η. As η increases from zero to the full value some elements of the sub-region may fail (when η is greater than ξ) and the *available strength reserve* R_i in the sub-region volume V_i

$$R_i = \frac{1}{V_i} \cdot \int_{V_i} (\xi - \eta) \mathrm{d}V$$

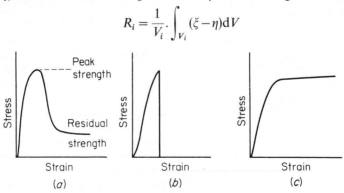

Figure 3.29   Stress-strain curves for soil

is not necessarily equal to the mean strength reserve

$$\bar{R}_i = (\bar{\xi}_i - \eta)$$

since the contribution to the reserve from the failed elements is only $(r \cdot \xi - \eta)$. Assuming that the elements behave independently the available strength reserve is given by

$$R_i = \frac{1}{V_i} \cdot \int_{V_i} \left\{ \int_{\xi > \eta} \xi \cdot g(\xi) d\xi + \int_{\xi < \eta} (r\xi) \cdot g(\xi) d\xi - \eta \right\} dV$$

$$= \bar{\xi} - \eta - (1 - \bar{r}) \cdot \frac{1}{V_i} \cdot \int_{V_i} \Phi(\eta) dV \qquad (3.36a)$$

with $\bar{r}$ the mean ratio of residual to peak strength and $\Phi(x)$ the first incomplete moment of x

$$\Phi(x) = \int_{-\infty}^{x} \xi \cdot g(\xi) d\xi$$

When $r = 0$, the stress-strain curve is that of *Figure 3.29b*, corresponding to a brittle material. This model of strength has been studied extensively by structural engineers and leads to the use of the lowest extreme value of ξ as the characteristic strength (Bolotin (1969)).

The common assumption in soil engineering is $r = 1$, with an elastic-plastic stress-strain curve as shown in *Figure 3.29c*. This leads to the extremely simple result, for both ξ and η being variables,

$$R_i = \bar{\xi}_i - \bar{\eta}_i \qquad (3.36b)$$

or, for the entire volume affected

$$\bar{R} = \bar{\xi} - \bar{\eta} \qquad (3.36c)$$

Anisotropy. If the strength of the soil element depends on the direction of orientation θ of the stress the pertinent pdf to use in equation 3.36 is the conditional pdf $g(\xi | \theta)$. It will usually be sufficient to assume that ξ is of the simple form

$$\xi_\theta = \xi_0 \cdot f(\theta, k_j)$$

where k_j are average values of parameters determined by testing specimens at different orientations θ, and ξ_0 is a random variable. For each sub-region the available strength reserve could be calculated from

$$R_i = \min_\theta \left[\bar{\xi}_\theta - \eta_\theta - (1 - \bar{r}) \frac{1}{V_i} \int_{V_i} \Phi(\eta_\theta) dV \right] \qquad (3.36d)$$

Discontinuities in the soil, such as fissures or joint-planes, form potential failure surfaces whose influence depends on the relative values of joint and matrix strength parameters and on the probability of occurrence of a critical joint-plane inclination. The conditional strength distribution will inevitably be bi-modal, with modes at the mean matrix strength and at the most critical inclination. For the decomposed granite of *Figure 3.9* the cdf of the plane-strain strength $\frac{1}{2}(\sigma_1 - \sigma_3)$, with vertical and horizontal major and minor principal stresses σ_1 and σ_3, is shown in *Figure 3.30* for low and high values of σ_3 and for two bearing angle directions δ of the plane of σ_3.

Repeated loadings. Since the live loads such as water pressures, earth-quake forces, or traffic loads, are random time processes the soil will be subjected to repeated loadings of various levels throughout the working life, and the soil response is not necessarily the same as would be predicted from a single application of a certain level of load. Cumulative damage (i.e. accumulation of permanent strain) occurs under repeated loadings and the amount of damage will depend on the number of repetitions of the various stress levels.

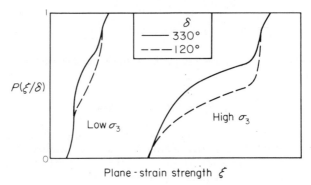

Figure 3.30   Plane-strain strength distribution for jointed soil

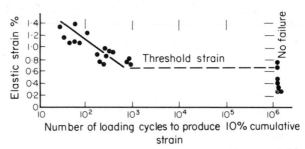

Figure 3.31   Cumulative damage curve for clay

Figure 3.31 illustrates the effects of repetition of fixed levels of applied strain on a clay soil (Heath *et al.* (1972)), from which it will be seen that there is a threshold level of applied strain (or stress) below which failure does not occur. Above the threshold level, a design would have a limited life based on the number of allowable load repetitions.

For random loads the order of application of the various levels will influence the cumulative damage curves, but it may be a reasonable design approximation to ignore the influence of order (Miner (1945)). If, during a time T, there are n_i applications of stress level η_i for which N_i applications would cause failure, failure under a combination of k different stress levels would be predicted if

$$\sum_{i=1}^{k} (n_i/N_i) \geqslant 1$$

Hence, if the pdf of η_i, $g(\eta_i|T)$ say, can be estimated the expected life before failure can also be estimated (Bolotin (1969)).

3.8.3 Stability analysis

Given perfect analytical methods and complete information, the strength reserves R_i of equation 3.36 could be calculated exactly for all sub-regions. With incomplete information R_i will be a random variable and hence there will be a probability $p_i = \text{Prob}(R_i \leqslant 0)$ of failure of the individual sub-regions, and the overall probability of failure p_N of the entire volume will be some function of the p_i values

$$p_N = f(p_i) \quad ; \quad i,\, 1 \text{ to } N$$

and an adequate design would be based on an acceptably low value of p_N.

If the sub-regions are assumed to act independently, that is the failure of one sub-region does not affect another sub-region $\text{Cov}(p_i, p_j) = 0$, then p_N can be calculated in some simple cases. Replacing the entire volume by a three-dimensional lattice with the original sub-regions regarded as the lattice vertices (the 'sites') or as the connections between vertices (the 'bonds'), two limiting cases are for the entire lattice to fail if a single site or bond fails (the series model) or for the entire lattice to fail if all sites or bonds fail (the parallel model). The overall failure probabilities are given by

$$\text{Series} \quad p_N = 1 - \prod_{i}^{N} (1 - p_i) \tag{3.37a}$$

$$\text{Parallel} \quad p_N = \prod_{i}^{N} (p_i) \tag{3.37b}$$

In actuality, failure will occur when only n out of N sub-regions fail (along a 'slip-surface', in 'plastic zones', etc.), these n sub-regions forming a mixed series-parallel system. Although p_N can always be bracketed by equation 3.37.

$$\prod (p_i) \leqslant p_N \leqslant 1 - \prod (1 - p_i) < \sum p_i$$

the limits will be too widely spaced to be of any practical use.

Still assuming independence and taking $p_i = p$ equal for all sub-regions, there is a critical value p_c below which $p_N = 0$, and an acceptable design could be based on $E(p) = p_c$. Unfortunately p_c is not unique, but depends on the type of lattice, on whether bond or site failure occurs, and on the shape of the lattice boundaries. Some estimates of p_c are given in *Table 3.15* for infinite lattices (Frisch *et al.* (1961)), and in *Table 3.16* for finite two-dimensional lattices (Dean (1963)).

Table 3.15 CRITICAL PROBABILITIES FOR INFINITE LATTICES

		Site	Bond
Two-dimensional	Triangular	0.49	0.34
	Square	0.58	0.49
	Hexagonal	0.69	0.64
Three-dimensional	Tetrahedral	0.44	0.39
	Simple cubic	0.32	0.25
	Close-packed hexagonal	0.20	0.12

Table 3.16 CRITICAL PROBABILITIES
FOR FINITE TWO-DIMENSIONAL LATTICES

Size of lattice	Critical probability (site)
6×96	0.71
12×48	0.59
24×24	0.57

The actual failure zone in nature is quite thin and hence analogous to a curved two-dimensional surface lattice. For unequal lattice probabilities p_i it may be possible to base a design on the critical probability for a triangular lattice, allowing for effects of boundary shape and size, if elastic-plastic behaviour as in *Figure 3.29c* can be assumed, using

$$E(p_i) = \bar{p} = \frac{1}{N} \sum (p_i) = p_c (\text{triangular})$$

This would be unsatisfactory for brittle soils with stress-strain curves as in *Figure 3.29a* or *b*, for here the lattice probabilities will inevitably be correlated. Failure of one sub-region will increase the failure probabilities in neighbouring sub-regions, since the excess loads must be transferred from the failed sub-region to maintain equilibrium. Critical probabilities p_c no longer exist, and the relation between p_N and $\bar{p}$ will be S-shaped in general as shown in *Figure 3.32*. It may prove satisfactory to base the design on a local average $\bar{p}_n$ taken over a limited highly stressed zone $V_n < V$

$$\bar{p}_n = \frac{1}{n} \sum (p_i) = p_c (\text{uncorrelated})$$

but this is speculative, since the theory of system reliability for correlated failure probabilities is far from complete at present.

Approximate analysis of stability. In the traditional method of stability design, the characteristic loads and strengths are modified by safety factors to give design values which are used in a deterministic manner, regarded as known constants.

For strengths, the characteristic values are currently chosen in a variety of ways; as the sample mean $\bar{x}$; as the smallest sample value $x_{(1)}$; as a quantile' value $x_p = (\bar{x} - s . u_p)$ with $p = 10\%$ (EAU (1971)). Since the sample mean $\bar{x}$ is the most efficient estimator, in comparison with $x_{(1)}$ or x_p, this is the recommended characteristic to use.

Safety factors. The strength has two components $\xi = c + \sigma . t$, where here σ is the normal direct stress, c the cohesion, and t the tangent of angle of shearing resistance. Similarly the load $\eta = \delta + \lambda$ say, where δ is the dead load and λ the live load component. The design response ζ will be some function of strength and load $\zeta = f(\xi, \eta)$. Taking the means as characteristics, central safety factors v are applied to give design values as follows

Design Strength $\quad \xi_d = \dfrac{1}{v_S} . \bar{\xi} = \dfrac{1}{v_c} . \bar{c} + \sigma_d . \dfrac{1}{v_t} . \bar{t}$ (3.38a)

Design Load $\quad \eta_d = v_L . \bar{\eta} = v_\delta . \bar{\delta} + v_\lambda . \bar{\lambda}$ (3.38b)

Design Response $\quad \zeta_d = \dfrac{1}{v_R} . f(\xi_d, \eta_d)$ (3.38c)

Table 3.17   TYPICAL SAFETY FACTORS

Strength	Cohesion	1.3–1.6
	$\tan \phi$	1.1–1.2
Load		
Dead. Soil weight, static water pressure		1.0
Live. Dynamic water pressure		1.2–1.5
Design Method		
Bearing capacity		2–3
Slope stability		1.0–1.5
Earth pressure		1.0–1.5
Anchor piles		1.5–2.0

Numerical values of these central safety factors as recommended by Codes of Practice or Regulations (Hansen (1967), EAU (1971)) are chosen by consensus of opinion on 'good' practice, and some typical recommended values are given in *Table 3.17*.

The intent of the strength and load factors v_S, v_L is to cater for variability, and of the response factor v_R to cater for bias and imprecision (c.f. equation 3.35). However, the use of values such as given in *Table 3.17* can lead to designs which are either over-cautious or unsafe in any particular case, since the actual variabilities are not taken into account directly. Better insight into the reliability of a design will always be obtained by re-interpreting the safety factors in terms of probabilities.

Semi-probabilistic design.   A simple by over-cautious approach can be based on the series reliability response model, assuming that failure will occur if any one of the sub-regions fails. From the pdf's of loads and strengths the pdf of response ζ can be calculated

$$P(\zeta_p) = \text{Prob}(\zeta < \zeta_p) = p^*$$

and, using the equation 3.38c the safety factor v can be associated with p^*

$$\zeta_p = \frac{1}{v} \cdot \zeta = \frac{1}{v} \cdot f(\bar{\xi}, \bar{\eta})$$

This value p^* will be an upper bound to the unknown probability p.

If elastic-plastic behaviour is assumed (as in *Figure 3.29c*) a lower bound to the probability p^{**} can be calculated based on the pdf's of the *means* $\bar{\xi}$, $\bar{\eta}$. Here, it is implied that failure occurs if the actual mean strength is less than $\bar{\xi}$ or the mean load greater than $\bar{\eta}$ (c.f. equation 3.36c) and hence p^{**} will depend on the sample sizes used in estimating $\bar{\xi}$ and $\bar{\eta}$. The actual failure probability is thus bracketed by

$$p^{**} < p < p^*$$

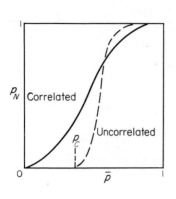

Figure 3.32   System reliability

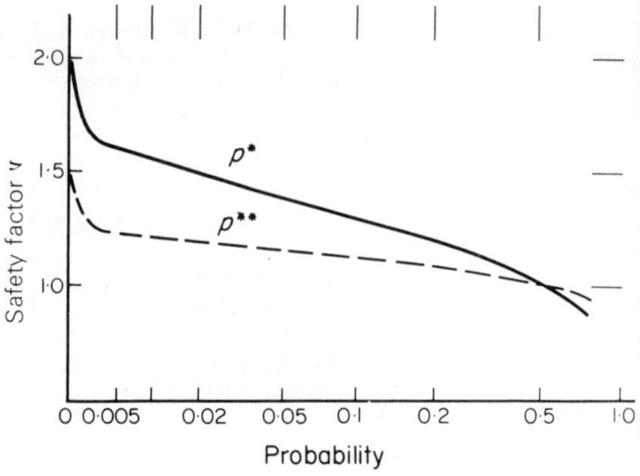

Figure 3.33   Probability bounds for slope stability

For a simple slope of height H and slope angle θ in 'uniform' soil (a single sub-region) of strength c, t, and density γ, the design equation is

$$H = \frac{c}{\gamma} \cdot f(\theta, t) = \frac{1}{v} \cdot \bar{H}$$

where $\bar{H}$ is the height calculated using mean values $\bar{c}$, $\bar{t}$, $\bar{\gamma}$. From a large number of tests on some residual soils (Lumb (1970)) the distributions of c and t are characterised by

	CV	$\sqrt{\beta_1}$	β_2
c	0.17	+0.50	2.73
t	0.06	+0.13	2.28

Taking a mean angle of shearing resistance of 35° and a slope angle of 60°, and ignoring the small variability of $\bar{\gamma}$ the probability p^* associated with the safety factor is shown in *Figure 3.33*, calculated by the numerical integration method of Section 3.3.

Assuming that $\bar{c}$, $\bar{t}$, were estimated from the results of n tests then the distribution parameters for the *means* are (from equation 3.8)

$$CV(\bar{x}) = \frac{1}{\sqrt{n}} \cdot CV(x)$$

$$\sqrt{\{\beta_1(\bar{x})\}} = \frac{1}{\sqrt{n}} \cdot \sqrt{\{\beta_1(x)\}}$$

$$\beta_2(\bar{x}) = 3 + \frac{\beta_2(x) - 3}{n}$$

and, taking a small sample of $n = 5$, the probabilities p^{**} can be calculated using the following values (*Figure 3.33*)

	CV	$\sqrt{\beta_1}$	β_2
$\bar{c}$	0.076	+0.22	2.95
$\bar{t}$	0.027	+0.06	2.86

For this slope and this soil, the failure probabilities are bracketed by

$$v = 1.3 \qquad 0.015 < p < 0.1$$
$$v = 1.5 \qquad 0.0005 < p < 0.02$$

Since in this case the pdf's of c and t are bounded Main Type 1 distributions, the pdf's of v are also bounded, giving

$$v^* = 2.05$$
$$v^{**} = 1.55$$

associated with zero probability of failure.

The lower bound method based on p^{**} has been applied by Müller (1965), Wu and Kraft (1967), and Cornell (1971) to a variety of stability problems. It should be noted that if the assumed pdf's are unbounded (e.g. Normal distributions) then the pdf of v is also unbounded, and a design value of v will be associated with a non-zero probability p^{**} or p^*.

3.8.4 Settlement analysis

Making the usual assumptions that shear strains and lateral strains can be disregarded, the settlement $s(x, y)$ at a point (x, y) on the ground surface is given by

$$s(x, y) = \int_0^H m . \Delta\eta . dz$$

where $m(x, y, z)$ is the compressibility, $\Delta\eta(x, y, z)$ the increase in vertical stress at depth z below the point, and $H(x, y)$ the thickness of compressible soil.

Replacing the soil volume by $n(x, y)$ thin horizontal layers of thickness h_i each of mean compressibility $\bar{m}_i$, with stress increase $\Delta\eta_i$ at the centre of each layer, then

$$s(x, y) = \sum_i^n \bar{m}_i . h_i . \Delta\eta_i$$

and substitution of the pdf's of $\bar{m}$ and $\Delta\eta$ would lead to the probability distribution $P(s|x, y)$ of the surface settlement. A detailed study of this approach has been made by Resendiz and Herrera (1969) using a Taylor's Series expansion for a rigid foundation of length L and width B with average bearing pressure η. Their final expressions can be written as

$$s(x, y) = \eta . (\rho_o + x . \psi_x + y . \psi_y) \qquad \begin{cases} -\tfrac{1}{2}L \leqslant x \leqslant \tfrac{1}{2}L \\ -\tfrac{1}{2}B \leqslant y \leqslant \tfrac{1}{2}B \end{cases}$$

where

$$E(\rho_o) = \sum_i^n \bar{m}_i . h_i . f_i$$

$$V(\rho_o) = \frac{A}{4LB} . \sum_i^n V(m_i) . (h_i)^2 . F_i$$

$$E(\psi_x) = E(\psi_y) = 0$$

$$V(\psi_x) = \frac{9A}{BL^3} . \sum_i^n V(m_i) . (h_i)^2 . F_i . a_i$$

$$V(\psi_y) = \frac{9A}{LB^3} . \sum_i^n V(m_i) . (h_i)^2 . F_i . b_i$$

with $V(m_i)$ the variance of m in the i-th layer and f_i, F_i, a_i, b_i, numerical factors depending on the depth ratio $z_i/(LB)^{\frac{1}{2}}$ of each layer and on the shape ratio L/B. The factor A is the cross-sectional area of the specimen used in measuring m, and the term $V(m)$. $A/(LB)$ is the variance of the mean in the layer of extent LB (c.f. equation 3.33b).

For a square foundation $V(\psi_x) = V(\psi_y) = \sigma^2$, say, and the total angle of tilt θ is

$$\theta = \eta.(\psi_x^2 + \psi_y^2)^{\frac{1}{2}} = \eta.\psi$$

and, assuming a Normal distribution for m, ψ follows the Rayleigh Distribution

$$P(\psi) = 1 - \exp\left\{-\frac{1}{2}\frac{\psi^2}{\sigma^2}\right\}$$

hence the mean tilt $\bar{\theta} = \eta.\sigma.(\pi/2)^{\frac{1}{2}}$

The effects of load variability can be included if the pdf's of η and m are known, by calculating the joint pdf's of

$$S_o = \eta.\rho_o$$
$$\theta_x = \eta.\psi_x$$
$$\theta_y = \eta.\psi_y$$

3.8.5   Decision theory

At all stages in a design from initial concept to final completion decisions have to be made (choice of characteristic values, type of foundations, construction methods, etc.) and the consequences of these decisions D_i depend on the unknown actual 'state of nature' θ. The possible states of nature θ_j which might exist will have prior probabilities $P(\theta_j)$ and, with current information X, the posterior probability $P(\theta_j|X)$ can be calculated using Bayes' Theorem, equation 3.3.

The consequence of making Decision D_i, assuming a state of nature θ_j can be expressed as a *loss function* $L_{ij} = L(D_i, \theta_j)$ representing the best judgement of actual loss incurred if θ_j turns out to be a false assessment of θ and, provided that L_{ij} and $P(\theta_j)$ can be estimated consistently, the 'best' or optimum decision to make, D^*, is that which minimises the *expected loss* L_i for all possible states of nature

$$L_i = \int L_{ij}.P(\theta|X).d\theta$$

Since it would not be practicable to consider the infinite exhaustive set of all possible decisions and states of nature only limited sub-sets can be taken into account, giving a modified expected loss function

$$L_i = \sum_j^k L_{ij}.P(\theta_j|X) \tag{3.39}$$

with optimum decision D^* that which minimises equation 3.39. In effect some possible but highly improbable states of nature are given zero prior probability and some possible but irrelevant decisions given zero loss function.

The whole decision process can be split up into a series of sequential decisions connected in a branching network or tree-diagram, with decision choices at each branch point. Calculating expected losses at each branch point and tracing a minimum expected loss path along the branches eliminates some of the (D_i, θ_j) members and leads eventually to a final decision (Raiffa and Schlaifer (1961), Jaynes (1963)). A full analysis involves considerable calculation but even a partial analysis can clarify the whole problem by picking out the critical decisions which dominate equation 3.39 and by indicating where current information is too vague and must be augmented by additional investigations.

Loss functions.   Choice of a suitable loss function L_{ij} is the main difficulty in applying decision theory to engineering design, for the consequences of a wrong decision $(\theta_j \neq \theta)$ range over a wide spectrum from a mild annoyance to a national disaster. The cost of repair work, compensation for property damage, or loss of availability of a project due to a failure, or extra construction costs due to ultra-cautious design, can be assessed fairly easily but other factors such as possible loss of life or potential loss of reputation arising from a failure are difficult to quantify accurately. Loss functions will rarely be calculable exactly, but usually the right order of losses can be estimated and this may be sufficient.

To take a simple example, suppose that a slope is to be designed and that the alternative states of nature are θ_1, the soil is fairly uniform, and θ_2, there is a critically inclined weak seam in the soil. The relative costs of construction, total costs including repairs of failures, and expected losses with a prior probability $P(\theta_2) = p$ are given below

Decision	State of Nature		Expected Loss
	θ_1	θ_2	
$\theta = \theta_1$	1	$(1+\alpha)$	$1+\alpha \cdot p$
$\theta = \theta_2$	$(1+\gamma)$	$(1+\beta)$	$1+\gamma+(\beta-\gamma)p$
Prior Probability	$(1-p)$	p	

If $p < \gamma/(\alpha+\gamma-\beta)$ then the optimum decision D^* is $\theta = \theta_1$, to assume uniform soil. For design of road cuttings in remote areas α, β, γ will be of the same order and unless p is very large, (close to 1) the design would be based on θ_1, ignoring the possibility of a weak seam. For design of an earth dam in a populated area where a failure could cause considerable loss of life then α will be very much greater than β or γ $(\gamma/\alpha < 10^{-6}$, possibly) and even a small value of p would lead to designing on $\theta = \theta_2$.

For a continuously variable state of nature, such as soil strength, the loss function will depend on $|\theta_j - \theta|$, the greater the deviation the greater the loss, but the significance of a loss L to a large organisation (a national authority or a major corporation) will commonly be far less than the significance to a small organisation (a minor contractor or a small consultancy). Bernouilli's principle of 'moral expectation' relates the significance of L to the current 'assets' C, the financial capital, reputation, etc., and leads to a loss function of the form $\log(1+L/C)$.

If C is very large this gives a loss function

$$L = k.|\theta_j-\theta|$$

and if C is small a loss function

$$L = \log\{1+k.|\theta_j-\theta|\}$$

Pre-posterior analysis. Suitable prior probabilities $P(\theta_j)$ can be chosen with the help of the coefficients of variation of *Tables 3.5, 3.10, or 3.12* and the pdf shape factors of *Figure 3.20*, or alternatively in a subjective manner based on experience (Savage (1954); Folayan *et al.* (1970)). The need for gathering additional information X can then be estimated by calculating the *pre-posterior* probability $P(\theta_j|X)$, along the lines of Section 3.5.3, before actually obtaining the additional information.

The cost of obtaining the information will depend on a sample size n (number of boreholes, number of test specimens, etc.) and the expected loss $L_{i,n}$ will be of the form

$$L_{i,n} = \sum L_{ij}.P(\theta_j|n)+L(n) \qquad (3.39a)$$

and an optimum sample size n^* found by minimising equation 3.39a.

This type of analysis has been applied to foundation design (Rosenblueth (1969), Turkstra (1970)) based on expected settlements. Assuming the soil compressibility to have variance $V(m)$ and the *mean* compressibility a variance $V(\bar{m})$, the optimum sample size depends on the ratio $\alpha = V(m)/V(\bar{m})$. Vague prior information corresponds to a small value of α and precise prior information to a large value of α. *Figure 3.34* (after Turkstra (1970)) shows typical results of the analysis, assuming a Normal distribution for m, with a loss function depending on the expected depth of foundation $E(d)$ and expected settlement $E(s)$.

$$L = 1+C_1.E(d)+C_2.E(s^2)$$

For vague prior information ($\alpha = 1$) the optimum sample size is about 35, while for precise prior information ($\alpha = 10$) the optimum sample size is zero.

In the early stages of the analysis the optimum decisions will depend essentially on the assumed prior probabilities, but as more information becomes available the posterior probabilities $P(\theta_j|X)$ will become more and more precise and eventually the choice of the loss functions L_{ij} will dominate the analysis.

Choice of design probability. In the final stage the ultimate decision will be a choice of dimension (foundation width, slope angle) associated with a lumped parameter representing the total uncertainties. This parameter could be the estimated failure probability or the safety factor of Section 3.8.3 associated with a probability. Optimum design would be based on minimising equation 3.39 with respect to this parameter p to give a design probability p^* (Langejan (1965))

$$\frac{\partial}{\partial p}\{L(p)\} = 0$$

As mentioned before, the consequences of a failure differ for different projects, and there is no reason to expect a single probability value (or safety factor) to be optimum for all projects or even for all constituent parts of a particular project. A design probability of failure as high as 0.1 or 0.5

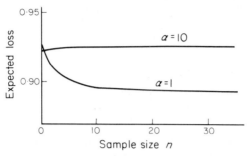

Figure 3.34   *Expected losses against sample size*

may be perfectly adequate when the cost of failure is limited to the cost of clearing away debris (as in open-cast mining), but a design probability as low as 10^{-6} may be too high if failure implies great loss of life (as with a deep cutting close to a populated area).

Absolute safety can never be guaranteed, nor can the probabilities be estimated exactly, but the use of decision theory and loss functions does allow the relative sensitivity of the design to various uncertainties to be calculated. If the analysis shows that the expected consequences of a very rare but disastrous event would dominate then the proposed design should be changed to reduce these consequences, whereas if the expected consequences of numerous minor failures dominate then the rare disaster need not be considered.

Laplace's comment that probability is nothing but common sense reduced to calculation is very pertinent. Decision theory, with statistical treatment of the uncertainties, is really a quantification of engineering judgement, a systematic method for calculating the 'calculated risk'. Much work remains to be done before the method can be applied in a routine manner to all soil problems, for some aspects are still controversial (e.g. the use of subjective prior probabilities) or ill-defined (e.g. the concept of failure itself), but once the random nature of loads and soil parameters is accepted then probabilities are unavoidable. The traditional deterministic approach to engineering, hiding all uncertainties under a blanketing safety factor, can be quite misleading when applied to soils.

BIBLIOGRAPHY

AGTERBERG, F. P. (1970). 'Autocorrelation functions in geology', *Geostatistics* (Ed. D. F. Merriam), Plenum Press, New York, 113–141.

BEATON, J. L. (1968). 'Statistical quality control in highway construction', *J. Construction Div., Am. Soc. C.E.*, **94**, CO 1, 1–15.

BJERRUM, L. (1967). 'Progressive failure in slopes of over-consolidated plastic clay', *J. Soil Mech. and Fdns. Div., Am. Soc. C.E.*, **93**, SM 5, 3–49.

BOLOTIN, V. V. (1969). *Statistical methods in structural mechanics*, Holden-Day Inc., San Francisco.

BOWDEN, D. C., and GRAYBILL, F. A. (1966). 'Confidence bands of uniform and proportional width for linear models', *J. Am. Statistical Assoc.*, **61**, 182–198.

BOX, G. E. P., and JENKINS, G. M. (1970). *Time series analysis, forecasting, and control*, Holden-Day Inc., San Francisco.

COHEN, A. C. (Jr.), (1961). 'Tables for maximum likelihood estimates: singly truncated and singly censored samples', *Technometrics*, **3**, 535–541.

CORNELL, C. A. (1971). 'First-order uncertainty analysis of soils deformation and stability', *Statistics and Probability in Civil Engineering* (Ed. P. Lumb), Hong Kong University Press, Hong Kong, 129–144.

DAVIES, O. L. (Ed.) (1967). *The design and analysis of industrial experiments*, 2nd ed., Oliver & Boyd, London.

DAVIS, F. J. (1953). 'Quality control of embankments', *Proc. 3rd Int. Conf. SMFE, Zürich*, **1**, 218–224.

DEAN, P. (1963). 'A new Monte Carlo method for percolation problems on a lattice', *Proc. Cambridge Phil. Soc.*, **59**, 397–411.

EAU (1971). 'Recommendations of the committee for waterfront structures', *EAU, 1970*, Wilhelm Ernst & Sohn, Berlin.

ELDERTON, W. P., and JOHNSON, N. L. (1969). *Systems of frequency curves*, Cambridge University Press, Cambridge.

EVANS, D. H. (1967). 'The application of numerical integration techniques to statistical tolerancing', *Technometrics*, **9**, 441–456.

EVANS, D. H. (1972). 'An application of numerical integration techniques to statistical tolerancing, III–General distributions', *Technometrics*, **14**, 23–35.

FISHER, R. A. (1953). 'Dispersion on a sphere', *Proc. Roy. Soc. London* (A), **217**, 295–305.

FOLAYAN, J. I., HOEG, K., and BENJAMIN, J. R. (1970). 'Decision theory applied to settlement predictions', *J. Soil Mech. and Fdns. Div., Am. Soc. C.E.*, **96**, SM 4, 1127–1141.

FRISCH, H. L., SONNENBLICK, E., VYSSOTSKY, V. A., and HAMMERSLEY, J. M. (1961). 'Critical percolation probabilities (site problem)', *Physical Review*, **124**, 1021–1022.

GRANT, E. L. (1964). *Statistical quality control*, 3rd ed., McGraw-Hill, New York.

GRUBBS, F. E. (1969). 'Procedures for detecting outlying observations in samples', *Technometrics*, **11**, 1–21.

GUMBEL, E. J. (1958). *Statistics of extremes*, Columbia University Press, Columbia.

HAMMERSLEY, J. M., and HANDSCOMB, D. C. (1964). *Monte Carlo methods*, Methuen, London.

HAMMITT, G. M. (1966). *Statistical analysis of data from a comparative laboratory test program sponsored by ACIL*, U.S. Army Eng. Waterways Exp. Station, Corps of Engineers Misc. Paper No. 4–785.

HANSEN, J. B. (1967). 'The philosophy of foundation design: design criteria, safety factors, and settlement limits', *Symp. Bearing Capacity and Settlement of Foundations*, Duke University, North Carolina, 9–13, 88.

HASSELBLAD, V. (1966). 'Estimation of parameters for a mixture of normal distributions', *Technometrics*, **8**, 431–444.

HEATH, D. L., SHENTON, M. J., SPARROW, R. W., and WATERS, J. M. (1972). 'Design of conventional track foundations', *Proc. I.C.E.*, **51**, 251–267.

HOLTZ, R. D., and KRIZEK, R. J. (1971). 'Statistical evaluation of soil test data', *Statistics and Probability in Civil Engineering* (Ed. P. Lumb), Hong Kong University Press, Hong Kong, 229–266.

JAYNES, E. T. (1963). 'New engineering applications of information theory', *Proc. 1st Symp. on engineering applications of random function theory and probability, Purdue*, J. Wiley & Sons, New York, 163–203.

JEFFREYS, H. (1961). *Theory of probability*, 3rd ed., Oxford University Press, Oxford.

JOHNSON, N. L., NIXON, E., AMOS, D. E., and PEARSON, E. S. (1963). 'Tables of percentage points of Pearson curves for given $\sqrt{\beta_1}$ and β_2 expressed in standard measure', *Biometrika*, **50**, 459–498.

KENDALL, M. G., and STUART, A. (1958). 'The advanced theory of statistics'; *Vol. 1: Distribution theory*. (1961), *Vol. 2: Inference and relationship*, (1966), *Vol. 3: Design and analysis, and time series*, Griffin, London.

KRAFT, L. M., and YEUNG, J. Y-H. (1973). 'Simulating deformation of statistically heterogeneous earth structures', *Geotechnical Engineering*.

KÜHN, S. H. (1971). 'Quality control in highway construction', *Statistics and Probability in Civil Engineering* (Ed. P. Lumb), Hong Kong University Press, Hong Kong, 287–312.

LANGEJAN, A. (1965). 'Some aspects of the safety factor in soil mechanics considered as a problem of probability', *Proc. 6th Int. Conf. SMFE, Montreal*, **2**, 500–502.

LUMB, P. (1966). 'The variability of natural soils', *Canadian Geotechnical J.*, **3**, 74–97.

LUMB, P. (1970). 'Safety factors and the probability distribution of strength', *Canadian Geotechnical J.*, **7**, 225–242.

LUMB, P. (1971a). 'Estimating the strength of jointed soils', *Proc. 1st Aust.–N.Z. Conf. on Geomechanics, Melbourne*, **1**, 175–179.

LUMB, P. (1971b). 'Precision and accuracy of soil tests', *Statistics and Probability in Civil Engineering* (Ed. P. Lumb), Hong Kong University Press, Hong Kong, 329–346.

LUMB, P., and HOLT, J. K. (1968). 'The undrained shear strength of a soft marine clay from Hong Kong', *Géotechnique*, **18**, 25–36.

MANDEL, J. (1964). *The statistical analysis of experimental data*, Interscience Publishers, New York.

MATHERON, G. (1965). *Les variables régionalisées et leur estimation*, Masson et cie., Paris.

MINER, M. A. (1945). 'Cumulative damage in fatigue', *J. Applied Mechanics*, **12**, 159–164.

MORSE, R. K. (1971). 'The importance of proper soil units for statistical analysis', *Statistics and Probability in Civil Engineering* (Ed. P. Lumb), Hong Kong University Press, Hong Kong, 347–358.

MÜLLER, R. A. (1965). 'Establishment of the values of the safety factors when calculating foundations for stability', *Soil Mech. and Foundation Eng.*, No. 3, 147–150.

PEARSON, E. S., and HARTLEY, H. O. (1966). *Biometrika tables for statisticians*, 3rd ed., Cambridge University Press, Cambridge.

PETTIT, R. A. (1967). 'Statistical analysis of density tests', *J. Highways Div., Am. Soc. C.E.*, **93**, HW 2, 37–51.

QUENOUILLE, M. H. (1949). 'Problems in plane sampling', *Annals Math. statistics*, **20**, 355–375.

RAIFFA, H., and SCHLAIFER, R. (1961). '*Applied statistical decision theory*, Harvard University Press, Cambridge, U.S.A.

RESENDIZ, D., and HERRERA, I. (1969). 'A probabilistic formulation of settlement controlled design', *Proc. 7th Int. Conf. SMFE, Mexico City*, **2**, 217–225.

ROSENBLUETH, E. (1969). 'Panel discussion, Main Session 2', *Proc. 7th Int. Conf. SMFE, Mexico City*, **3**, 230–233.

RRL (1967). 'Control of state of compaction in mass earthworks', *Road Research Laboratory Leaflet No. LF 27*.

SAVAGE, L. J. (1954). *The foundations of statistics*, J. Wiley & Sons, New York.

SCHULTZE, E. (1971). 'Frequency distributions and correlations of soil properties', *Statistics and Probability in Civil Engineering*, (Ed. P. Lumb), Hong Kong University Press, Hong Kong, 371–388.

SELBY, B. (1964). 'Girdle distribution on a sphere', *Biometrika*, **51**, 381–392.

SHAPIRO, S. S., and WILK, M. B. (1965). 'An analysis of variance test for normality (complete samples)', *Biometrika*, **52**, 591–611.

SHERWOOD, P. T. (1970). 'The reproducibility of the results of soil classification and compaction tests', *Road Research Laboratory Report LR 339*.

SMITH, T. W., and PRYSOCK, R. H. (1966). 'Discussion of Quality control of compacted earthwork by Turnbull *et al*', *J. Soil Mech. and Fdns. Div., Am. Soc. C.E.*, **92**, SM 5, 142–145.

TANG, W. H. (1971). 'A Bayesian evaluation of information for foundation engineering design', *Statistics and Probability in Civil Engineering* (Ed. P. Lumb), Hong Kong University Press, Hong Kong, 173–186.

TOCHER, K. D. (1963). *The art of simulation*, English Universities Press, London.

TURNBULL, W. J., COMPTON, J. R., and AHLVIN, R. G. (1966). 'Quality control of compacted earthwork', *J. Soil Mech. and Fnds. Div., Am. Soc. C.E.*, **92**, SM 1, 93–103.

TURKSTRA, C. J. (1970). 'Applications of Bayesian decision theory', *Structural Reliability and Codified Design* (Ed. N. C. Lind), Solid Mechanics Division Study No. 3, University of Waterloo, Waterloo, 49–71.

WHITTLE, P. (1954). 'On stationary processes in the plane', *Biometrika*, **41**, 434–449.

WHITTLE, P. (1962). 'Topographic correlation, power-law covariance functions, and diffusion', *Biometrika*, **49**, 305–314.

WILK, M. B., and GNANADESIKAN, R. (1968). 'Probability plotting methods for the analysis of data', *Biometrika*, **55**, 1–17.

WU, T. H., and KRAFT, L. M. (1967). 'The probability of foundation safety', *J. Soil Mech. and Fnds. Div., Am. Soc. C.E.*, **93**, SM 5, 213–231.

Chapter 4

Behaviour of Unsaturated Soils

B. G. Richards

4.1   INTRODUCTION

Soils can be considered to be composed of three phases, i.e. solid, liquid and gaseous. When the pore pressure in the liquid phase is positive (e.g. in soil below a water table), any gaseous phase present in the soil can only exist as trapped gas at a higher pressure than the ambient air pressure on the soil.

This gas will tend to diffuse out of the system and the soil will tend to reach a fully saturated condition where all the pore spaces are completely filled with water. Therefore, except for a few instances, as is the case with the rapid inundation of any unsaturated soil, positive pore pressures will cause the soil to become saturated. On the other hand when pore pressures become negative, the air–water interfaces tend to become concave and recede into soil pores, large pores may empty and air bubbles may not only form but also grow in size. Therefore, where negative pore pressures exist, the soil will tend to be unsaturated.

However, even up to significant values of negative pore pressure (as high as $10\,000\,\text{kN.m}^{-2}$) in many clay soils (Holmes (1955)), soils remain essentially saturated and, for practical purposes, their behaviour can still be assumed as being saturated. This has been defined as a 'quasi-saturated' state (Aitchison (1956)). In general, however, by measuring pore water pressure the soil can simply and conveniently be assessed as being saturated or unsaturated.

Current soil mechanics theory and practice have mainly been developed in those areas of North America and Europe where high water tables and freeze–thaw conditions are usual and soils are predominantly saturated or near saturated for many months of the year. Consequently the simplifying assumptions made are applicable only to saturated soils, composed of only the solid and liquid phases. Geometric relationships are therefore simple and as both phases are relatively incompressible, the soil can also be assumed to be incompressible under undrained conditions. Furthermore the effective stress law holds for practical purposes which can lead to simpler laboratory test procedures.

Since semi-arid to arid climatic conditions predominate in many of the newer developing nations such as Australia, the soils are usually unsaturated throughout the whole year and pose many problems to soil engineers. Erroneous predictions and incorrect decisions often result from the use of current practice, which is not applicable to unsaturated soils. A good example is the lack of appreciation of water flow through unsaturated soils. In many instances, granular layers, which may work as drainage layers in saturated soils, will actually lead to the wetting up of the unsaturated soil and cause serious heave and failure.

Although present day practice is as yet more or less unaffected by the research that has been carried out into the behaviour of unsaturated soils and the development of new theories and techniques, the prospects for the future are most promising, particularly in the field of road and airfield pavements, building foundations, retaining walls and natural and man-made slopes. Consequently, this chapter will briefly summarise the current experimental techniques and theories and will make little mention of the current practices.

4.2   RETENTION OF WATER BY SOILS

4.2.1   Definition of soil suction

When pore water pressures are positive, there is no conceptual difficulty in understanding the presence of water in soil. However, when pore water pressures become negative, such as in soil above a water table, the problem of understanding the mechanisms becomes much more difficult. In soil science, this problem is now treated thermodynamically, where the mechanisms of soil water retention need not be considered in developing a rational understanding and theory of behaviour.

When an unsaturated soil comes in contact with a pool of free water, *Figure 4.1*, it absorbs water; the soil exerts a suction, i.e. the soil suction, on the free water. This can be considered to be a convenient reference state. One way to quantitatively define this soil suction is to apply a suction of the same value to the pool of free water. When the suction on the free water is such that flow between soil and water ceases, then we have a quantitative measurement of soil suction. Using this concept, we can define soil suction as quoted by the International Society of Soil Science:

Soil suction (h) is that negative gauge pressure, relative to the external gas pressure on the soil water (normally atmosphere pressure), to which a pool of pure water at the same elevation and temperature must be subjected in order to be in equilibrium with the soil water.

The application of negative gauge pressure on the pool of free water is quite conceivable up to $100 \, kN \cdot m^{-2}$ suction, e.g. it can readily be applied by a standard vacuum pump. When negative gauge pressures or soil suctions exceed 1 atmosphere (values of over $100\,000 \, kN \cdot m^{-2}$ are often measured), the simple concept of negative gauge pressure is no longer valid. Effective suctions, however, can be applied to free water by thermal gradients (thermo-osmosis), electric fields (electro-osmosis), solute concentration differences (chemical or solute osmosis) or by elevating the reference gas pressure above

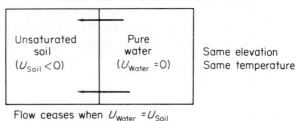

Flow ceases when $U_{Water} = U_{Soil}$

Figure 4.1   Diagrammatic representation for soil suction

atmospheric pressure. Although an actual negative gauge pressure is not used on the pool of free water, it is quite correct to define an equivalent or effective gauge pressure.

4.2.2   Definition of soil water potential

Soil suction can also be determined by the more flexible thermodynamic variable of potential, or in this particular case soil water potential. According to the definition quoted by the International Society of Soil Science:

> Soil water potential (φ) is the work done per unit quantity of pure water in order to transport reversibly and isothermally an infinitesimal quantity of water from a pool of pure water outside the adsorptive force fields at a specified elevation and at atmospheric pressure to the soil water (at the point under consideration).

In terms of the model in *Figure 4.1*, the pool of pure water has its potential lowered to that of the soil water. The concept of an equivalent soil suction which would produce the same lowering of the soil water potential as occurs in the soil now has a thermodynamic definition.

4.2.3   Mechanisms of soil water retention

While the mechanisms of soil water retention are not essential to the understanding of the behaviour of unsaturated soils, a brief mention of them will be made for completeness.

(a) Surface tension.   If soil is considered to be composed of inert soil particles, the pores formed between these particles can be regarded to act as capillary tubes. At air–water interfaces, the unequal attraction forces on water molecules give rise to the equation of capillarity in its simplest form;

$$h = \frac{2T}{r} \qquad (4.1)$$

where h = negative pressure in the water
               = soil suction;

T = surface tension of water;

r = radius of curvature of the meniscus.

The radius of curvature of menisci can be related to the radius of the capillary tube or soil pore, and, for practical purposes, both are equal. Therefore, at equilibrium, the higher the suction is, the smaller are the pores in which the air–water interfaces will occur. In fact, the capillarity theory provides simple geometric models to adequately simulate coarse-grained soils.

(b) Adsorption of water on clay minerals.   In most natural soils and particularly in expansive clays, the soil particles consist of a quite significant fraction of clay mineral particles, which cannot be considered as inert. The clay particles, with the possible exception of fractures surfaces, have negatively charged surfaces. In some clay minerals, e.g. montmorillonites, the particles are composed of molecular sheets or lattices separated by water layers of varying thickness and the internal surfaces are also negatively charged.

These negatively charged surfaces attract water, called adsorbed water, with an energy of attraction, which is a function of the distance of the water from the clay surfaces. Some of the various mechanisms suggested, are:

(1) *Hydrogen bonding.* The same bonds which form with the water can also form between the hydrogen ion (H^+) and the oxygen (O^{--}) or the hydroxyl (OH^-) ions which normally compose the surface of clay minerals.
(2) *Polar adsorption.* Since water molecules are not symmetrical, resulting in polarized charges, the positive side of water molecules can therefore be attracted and aligned towards the clay surfaces.
(3) *Cation adsorption.* The positively charged cations, e.g. Na^+, Ca^{++}, etc., which also adsorb water around them, are themselves attracted to the negatively charged clay surfaces.
(4) *London–Van der Waals secondary valence forces.* These are longer range forces arising from the electro-magnetic attraction between electric fields associated with the movements of electrons within their orbits.

(c) Osmotic imbibition.   As mentioned above, when cations are adsorbed onto the clay surfaces, the adsorbed ions and water constitute what is called the diffuse double layer. This adsorption of cations with the double layer can establish a solute concentration difference between the soil water beyond the double layer and that in the double layer. Therefore, the potential or soil suction of the water beyond the double layer must be such that this water is in equilibrium with the water within the double layer.

In the case of expanding lattice type clay minerals, e.g. montmorillonites, a change in the potential of the water outside the inter-lattice spacing and the double layer will upset the state of equilibrium. Hence the water, which will move in or out of the double layer in order to re-establish equilibrium, will alter the lattice spacing. An increase in potential will therefore cause an increase in the volume of clay particles, i.e. swelling. The osmotic pressure developed in the ideal case can be calculated by:

$$P_o = RT(C_c - C_o) \tag{4.2}$$

where

P_o = osmotic pressure
    = soil suction;

R = gas constant;

T = absolute temperature;

C_o = number of moles of ions/litre in free pore water;

C_c = number of moles of ions/litre at the mid-plane between particles.

(d) Combined effect on soils.   In soil, all the above mechanisms of water may apply, but their influence may differ from point to point within the soil. This is demonstrated in the schematic diagram in *Figure 4.2*. However, at equilibrium, the total potential, which is the net result of all these mechanisms, must be equal at all points in the soil.

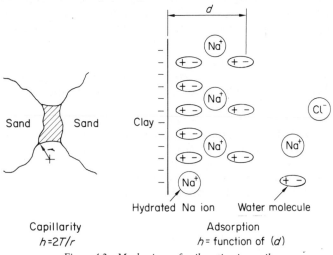

Capillarity
$h = 2T/r$

Adsorption
h = function of (d)

Figure 4.2   *Mechanisms of soil suction in a soil*

Another point, which can be brought out, is that the soil water need not have a negative gauge pressure in order to exist at a high soil suction. Probably the only mechanism mentioned above to create negative pressures is capillarity. Large values can only occur due to this mechanism when the air–water interface exists in very small pores, where the other mechanisms will begin to be dominant. Estimates of this negative pressure or tension in the soil water have been made, e.g. *Figure 4.3*, but they have not been verified and therefore should only be treated as a guide.

4.2.4   Components of soil water potential

While the mechanisms of soil water retention do not form useful components of the soil water potential, it is sometimes convenient to subdivide soil water potential into components on the basis of the forces responsible for the

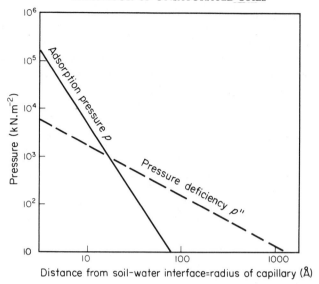

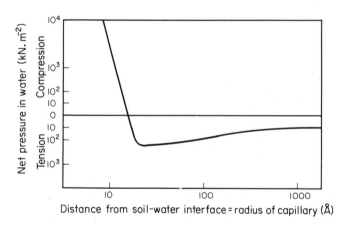

Figure 4.3 Pressures in soil water

differences in energy between the water in question and that of the reference state. These components are not necessarily related to the actual mechanisms.

(a) Matric potential (φ_m)   This results from the interaction of water with the adsorptive force fields emanating from solid surfaces. The equivalent soil suction component is the matric suction (h_m).

(b) Osmotic or solute potential (φ_s)   This results from the interaction of water with force fields emanating from dissolved substances. The equivalent soil suction component is the solute suction (h_s).

(c) Pressure potential (φ_p).   This results from the difference in energy of water caused by an external pressure other than that applied to the reference water.

(d) Gravitational potential (φ_z).   This results from the potential energy
water has due to its position in gravitational field relative to that of the
reference water. There is no equivalent soil suction component.

A schematic representation of these components is shown in *Figure 4.4*.
In general the equivalent soil suction for each component can be derived
from the soil water potential using the equation:

$$-\gamma_w gh = \varphi \qquad (4.3)$$

where γ_w = density of water

g = gravitational acceleration

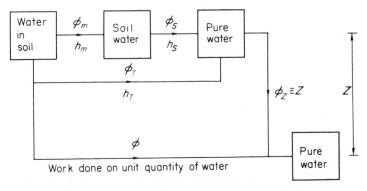

Figure 4.4   Diagrammatic representation of the components of soil suction

However, as φ is related usually to a fixed point in space, e.g. the water
table level, and h to the same elevation as the point in consideration, equation
4.3 is better expressed as:

$$\gamma_w g(z-h) = \varphi$$

or $h = z - \varphi/j_w g \qquad (4.4)$

where z is the height at which h is measured above datum for φ.

4.2.5   Soil suction as a quantitative variable

In treating soil water as a variable in the behaviour of unsaturated soil, we
require some quantitative expression of the state of the soil water. There are
basically only two forms of expressing quantitatively the soil water variable:

(1) as geometry of the soil water,   and
(2) as energy of the soil water.

The geometry of soil water can be expressed in various ways:

Gravimetric water content (w)

$$w = \frac{W_w - W_d}{W_d}$$

where W_w = wet weight of soil;

W_d = dry weight of soil.

Volumetric water content (θ)

$$\theta = \frac{V_w}{V_s}$$

where $\quad\quad\quad V_w$ = volume of water in soil;

$\quad\quad\quad\quad\quad\quad V_s$ = volume of soil.

Degree of saturation (S_v)

$$S_v = \frac{V_w}{V_v}$$

where $\quad\quad\quad V_v$ = volume of soil voids.

The energy state of the soil water can be expressed in many forms such as soil suction or soil water potential as discussed previously.

Both the water content and soil water potential, which are common forms of each type of expression are relatively easy to measure in the laboratory and are becoming easier to measure *in situ* in field applications. In fact, the soil water potential is, if anything, the easier to measure. In addition, it can be shown to represent the state of the soil water in soil much more effectively than water content.

(a) Wetness. For a simple description of a soil as being either wet or dry, water content cannot be used without a considerable knowledge of numerous other factors such as soil type, clay content, density, etc. For example, to describe an unknown soil as having a water content of 12% is meaningless. This is because if it were sand, the soil may be very wet, whereas for a clay it may be very dry.

On the other hand, it is more meaningful to say that a soil has a soil suction of 100 kN.m², because under Australian conditions, for example, such a soil would obviously be moist, even if we knew nothing else about the soil.

(b) Soil water flow. It is well established that the water flow through any porous media such as soil is controlled by soil water potential gradients as will be discussed later. Therefore, by measuring the soil water potential or soil suction, these potential gradients can be clearly established, and not only the direction of flow but also the flow rates estimated. For example, flow always takes place from high to low potentials or soil suctions as shown in *Figure 4.5*.

In a given soil, soil suction generally indicates high water content; however, this is not true when soil variability and changing soil types are considered. Therefore (*also shown in Figure 4.5*), water will generally flow from high to low water contents in uniform soil, e.g. sand or clay. When different soil zones are encountered such as in layered soil profiles, water will commonly flow from low water contents to high water contents. In general, therefore, not even the direction of flow can be predicted from water contents, let alone the rates of flow.

(c) Soil water variability. Soil in its natural state is an extremely variable material. Since soil suction, however, is primarily controlled by the soil environment and not by the soil itself, it tends to reach a uniform value in

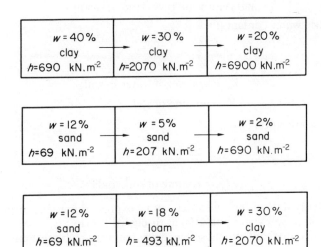

Figure 4.5   Examples of water flow in soils

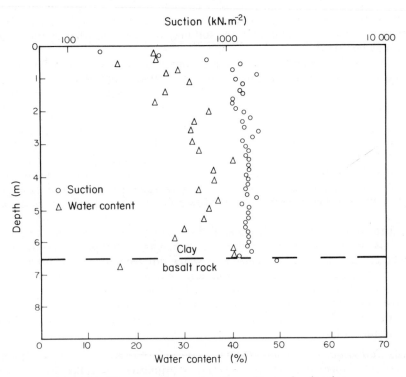

Figure 4.6   Water content and soil suction profiles in a basaltic clay

the soil through dissipation of any potential gradients, and its magnitude depends on the environment. The significant spatial and time variations in soil suction which actually occur in soil are reflections of changing environmental conditions, such as load and climate, and relevant soil characteristics, such as different soil zones. These variations are meaningful and can be used to analyse the previous history of the soil.

Water content, for a given soil suction, is highly sensitive to soil material variability such as variations in soil type, clay content, density, soil structure, etc. For example, typical variations in water contents in clay soils occurring over only a few centimetres at depths well before seasonal variations in soil moisture would be 30%±5% water content. A good example of the comparison between soil suction and water content measurements in a practical situation is shown in *Figure 4.6*.

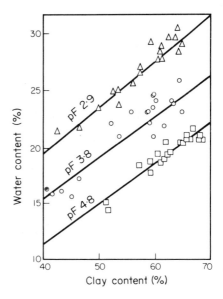

Figure 4.7 Water content—clay content relationships as function of soil suction in a red clay

Aitchison and Richards (1965) determined the relationship between clay content and water content at constant soil suction in the B-horizon of a Red-brown Earth. This soil was described as a moderately active clay with

| Liquid limit | 70–80% |
| -2μ fraction | 50–65% |

The results for three soil suctions controlled in a pressure membrane apparatus are shown in *Figure 4.7*. Included on this figure are best-fit correlations, resulting from statistical analyses, which showed that

$$N.(\Delta w)^2 = 74 \tag{4.5}$$

where N = number of samples required on each sampling occasion to prove a difference (Δw) of means of water content significant at a normal level of probability ($P = 0.05$). This suggests that if a difference in water content (Δw) of 1% is to be established with a probability of 0.05, then 74 samples must be taken at each location and time.

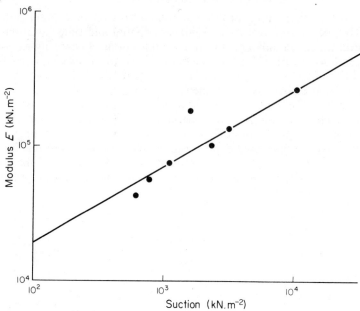

Figure 4.8 Soil modulus versus soil suction for a clay. Type M

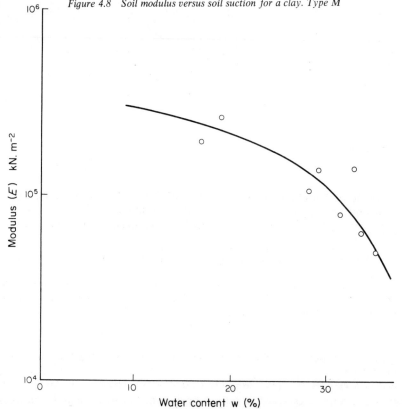

Figure 4.9 Soil modulus versus water content for a clay. Type M

Therefore it can be concluded that water content is not a unique variable of the soil and it is difficult to measure small changes in water content with time.

(d) Correlation of soil parameters with soil water variable. The correlation of any soil parameter such as shear strength and moduli with water content is generally poor unless other soil factors such as density and clay

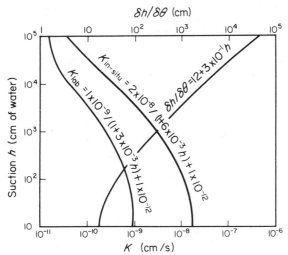

Figure 4.10   Permeability and $\partial h/\partial \theta$ versus soil suction. Type H

content are also considered. Similar correlations with soil suction are often better and certainly never worse than those made with water content. A good comparison of soil modulus versus soil suction and water content is shown in *Figures 4.8* and *4.9*.

As shown later, theoretical analyses of the transient behaviour of soils can only be made in terms of soil water suction, with which the relevant soil parameters can be most conveniently correlated. Other typical correlations are shown in *Figures 4.10* and *4.11*.

4.2.6   Measurement of soil suction

The many control and measurement techniques of soil suction, available for use in soil engineering applications, and the range of soil suction over which they will operate are shown in *Figure 4.12*.

Although some of these techniques, i.e. suction plate, pressure membrane, centrifuge, consolidation and vacuum desiccator only control the soil suction, they can be applied to measure soil suction on soil samples in the laboratory, by using what has been called the 'null-point' method (Coleman and Marsh (1961), Aitchison and Richards (1965)). Because of the less cumbersome and more accurate techniques now available, this method can no longer be recommended. These techniques have been reviewed elsewhere, (Croney *et al.*

124

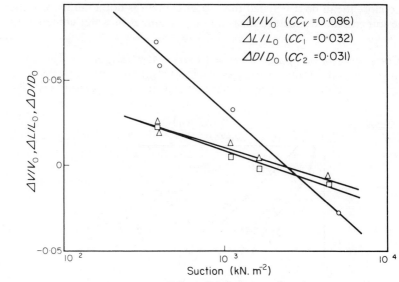

Figure 4.11 Volume changes versus soil suction

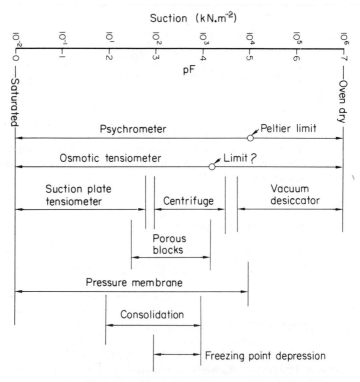

Figure 4.12 Soil suction measurement techniques and their useful range

(1952), Richards (1968), Coleman and Marsh (1961)) and will not be discussed here.

Of the direct measurement techniques, the osmotic tensiometer (Peck and Rabbage (1966)) and the freezing point depression method (Croney *et al.* (1952)) have not been completely successful and have not been accepted generally. The range of the tensiometer, which can be considered as only a variation of the piezometer, widely used for the measurement of positive pore pressures, is so limited that it has practically no value in most engineering applications.

Therefore, porous blocks, such as the gypsum block, and the psychrometer are probably the only ones worth considering. The gypsum resistance blocks were the only available means of measuring soil suction for many years. While they are now losing favour, the possible development of improved and cheaper thermal blocks warrants further discussion. The psychrometer is however emerging as the most promising and versatile technique and is receiving increasing acceptance in engineering applications.

4.2.7   Porous blocks

The water content of porous media can be remotely measured by some physical property of the soil that is sensitive to water content:

(1) Electrical resistivity or conductance.
(2) Dielectric constant or capacitance.
(3) Thermal conductivity.

Soil suction, usually the matrix suction, can also be estimated from the measurement of the soil water content and correlated with the above soil properties. However, there is no unique relationship between water content and soil suction as discussed previously and variations of up to $\pm 5\%$ water content are often observed in soil over a few centimetres at the same soil suction. A similar variation can be expected between soil suction and the soil property, e.g. soil resistivity.

The problem of soil variability can be overcome by measuring the physical property in a porous block of carefully controlled and known moisture characteristics embedded in the soil. From *Figure 4.13*, it can be seen that at equilibrium:

$$h_{block} = h_{soil}$$

and
$$w_{block} \neq w_{soil}$$

where h = soil suction (assumed to be matrix suction); and w = water content.

Typical soil suction-water content relationships are shown in *Figure 4.14*, indicating in these particular materials the useful range of soil suction (i.e. for a sensitive change in water content) for:

Plaster	$10-1000 \text{ kN} . \text{m}^{-2}$
Alumina cement	$100-100\,000 \text{ kN} . \text{m}^{-2}$

A wide range of porous materials have been used in practice, e.g. nylon, fibreglass, gypsum plaster, clay ceramics and sintered glass or metal. Gypsum plaster has been the most widely used in the past, as block resistance is very sensitive to dissolved solutes in the soil water. Being slightly soluble in water it tends to maintain a constant concentration of solutes in solution, thereby providing a buffering action. This has been shown to be reasonably effective up to about 0.07% total soluble salts, a figure, however, commonly exceeded in the drier areas of the world.

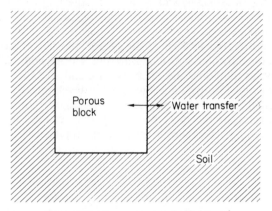

Figure 4.13   Porous block embedded in soil

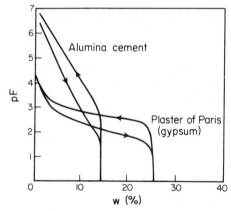

Figure 4.14   Soil suction—water content relationships for gypsum plaster and alumina cement

A typical layout of a gypsum block is shown in *Figure 4.15*. A.C. or oscillating current must be used to prevent polarisation of the soil water. Resistance can be relatively simply measured with modern a.c. resistance bridges. Typical calibration curves are shown in *Figure 4.16*.

Capacitance measurements have not been commonly used due to the difficulty experienced in the past of making stable and reliable measurements in the field. The measurements made have suggested a lower sensitivity to dissolved solutes in the soil water, but this needs further verification.

Thermal conductivity has also received little attention in the past, but with modern diodes, measurements are now cheap and relatively easy (Phene *et al.* (1971)).

Thermal blocks have shown little sensitivity to dissolved salts in the soil water or to the ambient temperature, and are therefore the most promising form of porous blocks currently available.

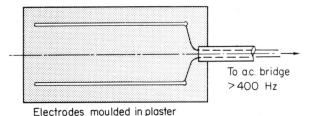

To a.c. bridge
> 400 Hz

Electrodes moulded in plaster

Figure 4.15   Schematic diagram of gypsum block

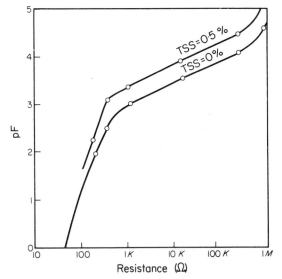

Figure 4.16   Calibration curve for gypsum block

The main disadvantages of porous blocks can be summarised as follows.

(a) Slow response times.   In clay soils of low permeability, effective response times may be of the order of many months. Therefore, the blocks are not generally suitable for transient measurements.

(b) Contact between soil and block.   Poor contact between soil and block can increase response times significantly and cause moisture flow to take place in the vapour phase. Vapour flow can be very slow and very sensitive to temperature. Since it will also result in equilibrium of total suctions (not matrix suctions), inaccurate reading is obtained. It should also be noted that good initial contact with expansive clay will not necessarily be maintained due to the inevitable volume changes, which subsequently take place.

(c) Physico-chemical effects.   Many fine-grained soils, e.g. heavy clays, act as semi-permeable membranes (Bolt and Lagerwerff (1964), Aitchison

and Richards (1969)) and the water transfer between the porous block and soil undergoes a change in chemical composition, resulting in equilibrium being established in the block corresponding to a soil suction between the total and matrix suctions.

(d) Hysteresis.   The hysteresis in the moisture characteristics of porous materials can introduce calibration difficulties and therefore serious errors. Hysteresis is a function of the material, but typical errors due to hysteresis are of the order of ±0.25 pF unit.

4.2.8   Psychrometric technique

At equilibrium, the vapour pressure, p, of the water vapour in the air above a soil surface is given by the thermodynamic equation:

$$h_t = \frac{RT}{\gamma_w g} \log_e \frac{p}{p_o} \tag{4.6}$$

where h_t = total soil suction;

   R = gas constant;

   T = temperature °K;

   γ_w = density of water;

   g = gravitational constant;

   p_o = vapour pressure over free pure water.

Because of temperature problems, it has not been practical to measure p separately over the soil and p_o over free pure water. In fact, the change in p

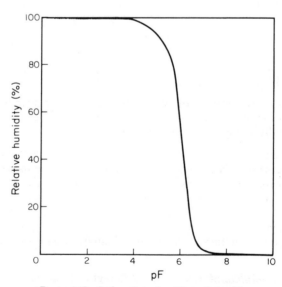

Figure 4.17   Soil suction—humidity relationship

per °C is about 80 times its change per atmosphere soil suction. However, it is now possible to measure the ratio, p/p_o (i.e. the relative humidity, H) simply and accurately with the psychrometric technique. With this technique the wet bulb temperature depression is measured by using wet and dry bulb thermometers and correlating it with humidity. Thus:

$$h_t = \frac{RT}{\gamma_w g} \log_e H/100 \qquad (4.7)$$

where $\quad H = p/p_o \times 100 =$ relative humidity % at 293 °K;

$\qquad PF = \log_{10} h_t = 6.502 + \log_{10}(2 - \log_{10} H)$

which is plotted in *Figure 4.17* and clearly shows the accuracy of measurement of humidity required in the usual soil suction range of 0–5 atmospheres.

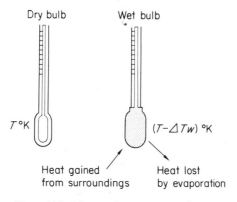

Figure 4.18   *Mercury thermometer psychrometer*

Figure 4.18 shows a typical mercury thermometer psychrometer. Due to the heat loss through the evaporation of water from the wet bulb, the temperature of the wet bulb will be lowered. If no heat is gained from the surroundings, the maximum lowering of the temperature of the wet bulb occurs when the vapour pressure of the free water on the wet bulb is lowered to that of the ambient humid atmosphere and the soil water when evaporation ceases. The expression for the rate of lowering of the wet bulb temperature is:

$$C\frac{dT}{dt} = -Lq \qquad (4.8)$$

where $\quad C =$ heat capacity of the wet bulb;

$\qquad T =$ temperature °K;

$\qquad t =$ time;

$\qquad L =$ latent heat of evaporation;

$\qquad q =$ rate of evaporation (function of T and h_t).

130 BEHAVIOUR OF UNSATURATED SOILS

Solving this equation gives the maximum wet bulb temperature depression, τ, where

$$\tau = \frac{gMT}{L} \times h_t \qquad (4.9)$$

and M = molecular weight of water.

In practice, however, heat is gained by the wet bulb in a number of ways, e.g.

(1) conduction through air (H_a),
(2) conduction through thermometer (H_w),
(3) net radiation from surroundings (H_r), and
(4) convection through air (H_c).

Equation 4.8 can be rewritten (Peck (1962)) as:

$$C \frac{dT}{dt} = H_a + H_w + H_r + H_c - L_q \qquad (4.10)$$

$$\Delta Tw = 41.8 \times 10^{-6} \, ^\circ K/kN.m^{-2} \qquad (4.11)$$

As can be seen, the very accurate temperature measurements required can be made with more or less conventional psychrometers, such as the thermistor psychrometer (Richards (1965)) or the droplet thermocouple psychrometer (Richards and Ogata (1958)). With both these forms of psychrometer, it is necessary to remove the psychrometer from the soil and place a water drop on the wet bulb. This not only adds an appreciable amount of water to the system but, due to the time required to establish equilibrium, any temperature drifts in the system are undetectable. Furthermore, it is not suitable for *in situ* use in soils, where it cannot be repeatedly removed from the soil. However, due to the large size of the water droplet, this type of psychrometer has the advantage that it maintains a steady output for a considerable time.

In order to overcome the above disadvantages, the Peltier psychrometer *(Figure 4.19)* has been developed (Spanner (1951), Richards (1968), Rawlins and Dalton (1967)). This psychrometer can be inserted in the soil and readings

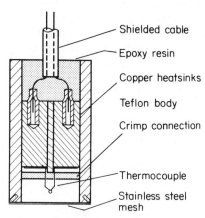

Figure 4.19 Peltier psychrometer

taken when required without removing it from the soil. It is therefore suitable for *in situ* use.

The basic principle of operation of the Peltier psychrometer consists of three steps:

(1) A measurement is taken of the thermocouple output corresponding to the dry bulb temperature, $T_i n$ μV.
(2) Water is condensed on the measuring junction immediately prior to making the wet bulb measurement by passing current through it and cooling it below the dew point. This forms a wet bulb.
(3) A second measurement is then taken of the thermocouple output which now corresponds to the wet bulb temperature, $T - \Delta T_w$ in μV.

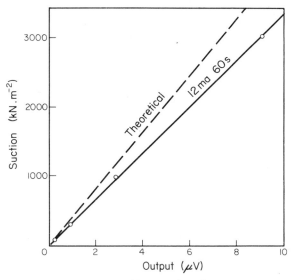

Figure 4:20   Calibration of psychrometer

In this case, the wet bulb depression ΔT_w is obtained by subtracting the output measured prior to wetting the junction from that measured immediately after. Because the junction is wetted only briefly and the quantity of water condensed on the thermocouple junction is small. The time for temperature drift is short, but the temperature unfortunately remains steady only for a brief time following wetting.

(a) Calibration.   With Chromel P–Constantan wires 0.50 mm dia. as shown in *Figure 4.19*, the theoretical calibration given by equation 4.11 becomes:

$$\Delta E = 0.00242 \ \mu V/kN.m^{-2} \tag{4.12}$$

The corresponding experimental calibration curve in *Figure 4.20*, obtained with standard NaCl solutions, gives good agreement, viz.

$$\Delta E = 0.00301 \ \mu V/kN.m^{-2} \tag{4.13}$$

(b) Response time.   The response time is very important in measuring transient conditions. As the air cavities in the soil in which the psychrometer is located is very small and only small amounts of water vapour transfer are necessary, response times can be theoretically shown (Peck (1968)) to be relatively short in most soils. Experimental measurements, made in the laboratory *(Figure 4.21)* correspond with these theoretical estimates and suggest that response times vary from a few hours at 100 atmospheres to about 14 days at 1 atmosphere.

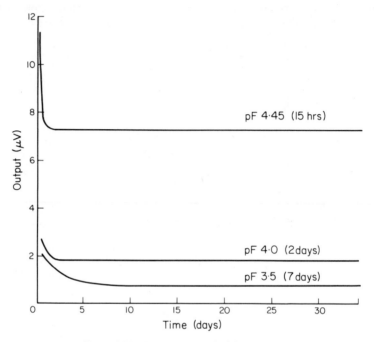

Figure 4.21   Response times for laboratory psychrometer

In laboratory use, response times in excess of one day is often inconvenient. The humidity in temperature controlled rooms is within the range of 50–70% humidity and a 24-hour reading can be used for practical purposes as all the response curves have a similar shape. An example of a 24-hour calibration curve is shown in *Figure 4.22.*

(c) Temperature effects.   Temperature effects fall into two types.

(i) Non-uniform temperatures.   As the ratio p/p_o must be the same for the soil and wet bulb, and p is a constant throughout the system at equilibrium, p_o and therefore T must also be the same. If however, a temperature difference exists between soil and wet bulb (T_{error}), it can be shown (Richards (1965)) that at 293 °K the error in maximum wet bulb depression (τ_{error}) is:

$$\tau_{error} = 19\,T_{error} \qquad\qquad (4.14)$$

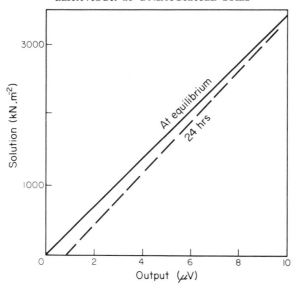

Figure 4.22   Twenty-four hour calibration curve for laboratory psychrometer

For example, when

$$h = 34.5 \text{ kN} . \text{m}^{-2}$$

$$\tau = 0.076 \,^{\circ}\text{K}$$

therefore $T_{\text{error}} \ll 0.004 \,^{\circ}\text{K}$

Hence temperature control in the laboratory must therefore be much better than $\pm 0.001 \,^{\circ}\text{K}$. The requirements for *in situ* use in the field will be discussed later.

(ii) Ambient temperature.   As both humidity and probe efficiency change with the ambient temperature, the calibration will also change with temperature. Experimental evidence suggests that at temperature, T, the soil suction corrected to $293 \,^{\circ}\text{K}$ is:

$$h_{293\,^{\circ}\text{K}} = \frac{h_T}{(0.027T - 6.91)} \tag{4.15}$$

where h_T is the apparent soil suction at temperature $T^{\circ}\text{K}$.

Thus the effect of ambient temperature on calibration is 2.7% per $^{\circ}\text{K}$.

(d) Peltier cooling.   When the cooling current, i, is passed, two effects occur:

(1) Joule heating of wires $(= i^2 R)$, and
(2) Peltier cooling of junction $(= Pi/_J)$.

With these two terms included in the heat balance equation (4.10), a quadratic in i is obtained, which can be solved for the current giving maximum cooling. However, this optimum current is best determined experimentally.

(e) Heat sinks.   When the measuring junction is cooled, the current, *i*, also generates heat in the reference junction formed at the heat sinks. The theory by Peck (1968) suggests that, for a temperature rise of 1% of ΔT_w, the heat sinks should be approximately 15 cc, which explains the use of the large heat sinks in the *Figure 4.19*. Large heat sinks also buffer heat transfer down the lead wires and help to maintain the junction at the same temperature as the soil.

(f) Use of psychrometers in situ.   Because of their sensitivity to temperature, psychrometers have until recently been used primarily in the laboratory, where temperature was controlled to $\pm 0.001\,°K$ or better. However with proper design, the major effects of temperature fluctuations can be eliminated (Rawlins and Dalton (1967), Richards (1968) and (1971), Lang (1968)).

These features are:

(1) Peltier cooling is used to compensate for temperature drifts.
(2) The measuring junction is very close to and surrounded by as great an area of soil as possible (Peck (1968)).
(3) Heat flow down lead wires is minimised by heat sinks and small thermocouple wires.
(4) An independent measurement of temperature is made to correct the soil suction reading.

In summary, the advantages and disadvantages of the psychrometer are as follows:

Advantages
(1) Cheap and easy to use.
(2) Measures total soil suction absolutely, i.e. there is no doubt as to what is measured.
(3) Good response times (normally less than 1–2 days).
(4) Covers the range of soil suction encountered in practice.
(5) Not affected by soil salinity.
(6) No apparent long term drift.

Disadvantages
(1) Calibration.
(2) Inaccurate where temperatures are changing rapidly.
(3) Calibration changes with large variations in ambient air pressure, but not significantly with normal atmospheric pressure changes.
(4) Corrosion of thermocouple wires can be a problem if coating breaks down.

4.3   FLOW OF WATER THROUGH SOILS

4.3.1   Principles of flow

It is now well established that moisture moves through soil under pressure or more correctly, potential gradients, having first been suggested by Darcy in 1856 for saturated filter beds and by Buckingham (1907) for unsaturated soils. The velocity and rate of moisture flow depends not only on the magnitude of the potential gradient, but also on the capillary flow properties of the soil as shown by the following equations:

$$v = K \frac{d\phi}{dl} \qquad (4.16)$$

or
$$Q = KA \frac{d\phi}{dl} \qquad (4.17)$$

where v = macroscopic velocity of flow;

K = coefficient of permeability;

$\dfrac{d\phi}{dl}$ = potential gradient;

ϕ = potential;

l = length along flow path;

Q = volumetric rate of flow (vol./unit time);

A = cross-sectional area.

For saturated soils, equation 4.16 is the well-known Darcy's law in which K, the permeability, is a constant for a given soil in a given condition. Philip (1958) using the Navier-Stokes equation, has shown that Darcy's law does not hold rigorously for all saturated flow conditions. The limitations to Darcy's law are due to *inertia effects*, and *unsteady flow conditions*.

The inertia effects cause a departure from Darcy's law when the Reynold's number exceeds 1 to 5, a condition rarely found naturally in soils. This departure is not due to the turbulence of the microflow in the pores as previously thought.

The second limitation has been found to be negligible except in rapidly changing systems. The early stages of infiltration may be subject to this limitation, but in most other circumstances of moisture flow in agriculture and engineering practice, any errors introduced by the use of Darcy's law may be neglected.

As soils become unsaturated, air replaces water in the pores and/or the pores diminish in size, impeding water flow in the liquid state. This decrease in the capillary flow property or permeability is due to three factors:

(1) The total cross-sectional area available for flow decreases with the volume of the water filled pores;
(2) From the capillary theory (Haines (1927)) it has been shown that:

$$r = \frac{2T}{\gamma_w h} \qquad (4.18)$$

where r = radius of curvature of air water interface;

$\equiv$ radius of pore;

T = surface tension;

γ_w = density of water;

h = soil suction.

At high moisture contents or low values of h, r is large. This means that the largest pores drain first, leaving the smaller pores filled. Poiseuille's equation for streamline flow in a narrow tube shows the relation between the velocity of flow and the radius of the tube:

$$v = \frac{r^2}{8v}\frac{d\phi}{dl} \qquad (4.19)$$

where v = radius of tube;

   v = viscosity;

(cf. equation 4.16),

$$v = -K\frac{d\phi}{dl}.$$

As the larger pores drain, the effective radius of the remaining pores must decrease, causing a reduction in the effective velocity of flow through the pores, which causes a further reduction in the permeability of the soil.

(3) As the volume of air-filled pores increases, more isolated water-filled pores can be expected. These contribute nothing to the liquid flow and further reduce the permeability of the soil.

In unsaturated soils, permeability is therefore no longer a constant independent of the water pressure, but a function of the volume of water-filled pores and their effective radii. Consequently, experimental relationships can and have been obtained between the unsaturated permeability and pore-space saturation or water content. Richards (1931), and Childs and Collis-George (1950) were among the first to show that the equation 4.16 applies to water flow in unsaturated soils when K is a function of the water content. As the water content of soils is some function of the soil suction, K can also be considered a function of soil suction, i.e. $K(h)$.

4.3.2   Theoretical prediction of $K(h)$

Marshall (1958), using Poiseuille's equation (1958) produced the relationship for saturated permeability:

$$K = \frac{n^2 m^{-2}}{\gamma v}[r_1^2 + 3r_2^2 + 5r_3^2 + \ldots + (2m-1)r_m^2] \qquad (4.20)$$

where n = porosity;

      m = integral number depending on degree of accuracy required;

   r_1 to r_m = pore radii;

and $r_1 > r_2 > \ldots > r_m$.

Equation 4.20 takes into account variations in pore sizes, which enabled Marshall to extend the equation to unsaturated soils. In the case of non-swelling soils, equation 4.18 can be rewritten as:

$$r = \frac{2T}{h} \tag{4.21}$$

giving $\qquad r^2 = 2.25 \times 10^{-2}h^2$ at $20\,°C$

where $h =$ soil suction in cm water, and

$$K = 2.7.10^2 n^2 m^{-2}[h_1^{-2} + 3h_2^{-2} + 5h_3^{-2} + \ldots + (2m-1)h_m^{-2}] \tag{4.22}$$

For the unsaturated case, where the water-filled porosity $= (m'/m)n$ and where $m' < m$, an estimate of $K(h)$ can be made, i.e.

$$K(h_{m'}) = 2.7.10^2 n^2 . (m')^{-2} \times$$
$$\times \; [h_{m'}^{-2} + 3h_{m'+1}^{-2} + 5h_{m'+2}^{-2} + \ldots + (2m-2m'+1)h_m^{-2}] \tag{4.23}$$

$K(h)$ can therefore be calculated from a plot of n versus h.

In practice, it has been found that for non-swelling soils, the function $K(h)$ from equation 4.23 usually differs from experimental curves. By matching the theoretical value to the experimental value at one point, i.e. at one soil suction, through the introduction of an empirical factor (a constant), good agreement over the whole suction range is, however, generally possible.

In swelling soils, the simpler capillary theory, used above, is no longer relevant, apart from the fact that the determination of n–h relationships is much more difficult. Therefore, the permeability function $K(h)$ for most natural unsaturated soils should be determined experimentally.

4.3.3   Measurement of $K(h)$

(a) Constant-head permeability test.   This technique is probably the most widely used for measuring permeability in saturated soils. Gurr (1962) and others have modified this technique so that the moisture flow is measured through an unsaturated soil sample under negative pore pressure or soil suction gradients. Not only is this technique limited to soil suctions less than 1 atmosphere, but difficulties are encountered in measuring the flow in soils of low permeability, e.g. clay soils. The soil suction range can be increased by raising the ambient air pressure on the sample in the same way as in the pressure membrane technique. The difficulties in doing this have prevented the widespread use of this technique.

(b) Transient flow in soil columns.   The solution of equation 4.16 for a known flow in a long column of unsaturated soil has also been used to determine the permeability function, $K(h)$. The suction gradients along the column are measured with tensiometers. With this method, the relationship between permeability and suction can be determined in one test over the suction range 0 to near 1 atmosphere. However, soil suctions of 1 atmosphere or more cannot be obtained and measured, unless the ambient pressure is raised and this has also met with many difficulties.

(c) Consolidation test.   The consolidation test is often used to determine the relationship between permeability of saturated soils and the voids ratio or water content. In clay soils of low permeability, difficulty is always

experienced in measuring very small flows, which can be insignificant compared with errors such as evaporation, leakage, side effects, etc. In these
soils, it is generally accepted that determinations of permeability of saturated
soils from the consolidation test are preferable.

In many fine-grained soils, the voids remain practically saturated to very
high soil suctions, often in excess of 100 atmospheres (Holmes (1955)).
Therefore, the effective stress laws are an approximation in these cases,
particularly the equivalency of applied load and soil suction in controlling
soil volume. As a result, the consolidation test can be used to determine the
relationship between permeability and soil suction. However, the technique
has two limitations:

(1) At the applied load in the test, all the voids are filled with water whereas
in the natural soil at the equivalent soil suction, some pores may
contain air. This is particularly so in structured soils, where the
effect can be most significant. The permeability so determined tends
to be higher than the true value at that suction.
(2) In most natural soils, the equivalence between applied load and soil
suction increasingly breaks down at the higher suctions. The suction
at which the permeability is measured consequently tends to be less
than the applied load used in the test.

These two sources of error act in the same direction, causing the
permeability as measured to be higher than the true value. This was clearly
shown by the experimental data from the consolidation and the pressure
plate outflow techniques (Richards (1965)).

(d) The pressure plate outflow technique.   The pressure plate outflow
technique was first developed by Gardner (1956) and has since been modified
and improved (Miller and Elrick (1958), Rytema (1959), Richards and
Richards (1962), Kunze and Kirkham (1962), Richards (1965)). This technique, which is very similar to the consolidation technique, is satisfactory for
all soil types, including the relatively impermeable clays, which are typical
of many Australian soils. It can also be used over the wide range of soil
suction found in Australia.

The outflow from a pressure plate is measured when a small suction
gradient is created by increasing the ambient air pressure. This has the
advantage over the consolidation technique in that

(1) The change in suction always equals the change in ambient air pressure.
(2) Side effects are unimportant.
(3) The flow parameter, $\partial h/\partial \theta(h)$ as discussed in Section 4.4 is also
determined.

The disadvantages are that:

(1) The volume of outflow is more difficult to measure than the volume or
height of sample.
(2) Poor sample contact on membrane or plate smearing of sample surface
and the low permeability of membranes or plates have a significant

effect on the flow conditions. Neglecting these factors may cause significant errors (Kunze and Kirkham (1962), Richards (1965)).

Typical results of $K(h)$ obtained by this method are shown in *Figure 4.10*.

(e) Field measurement.   The most serious criticism of the above methods is that the permeability is determined on laboratory samples and not *in situ* in the field. Not only does sample disturbance have an effect but sample size can be very important, especially where structural effects are evident in the soil, e.g. crack patterns (Richards (1968)).

Obviously the ultimate method of determining flow parameters must be a field technique. No such technique has been developed at this stage, but one *in situ* method, used successfully in one investigation, can be suggested (Aitchison *et al.* (1965)).

This method involves measuring soil suction *in situ* at a network of points spaced about the point of interest at different times. From the rate of change of soil suction and the spatial soil suction distribution, estimates of $K(h)$ can be made. Typical results are also shown in *Figure 4.10*, indicating the large variations which are possible between laboratory and field determinations.

The use of this method is relatively simple in uniform isotropic, non-structured soils. Variations of the method have in fact been applied in soil columns, tanks and natural profiles. To obtain permeabilities over a wide range of soil suction, artificial wetting or drying may be necessary. Care must be taken to ensure slow uniform changes as sharp gradients should be avoided. However, this condition is easily met under impervious surfaces such as road pavements or concrete slabs.

Other factors, which must be considered are: soil structure, particularly fissures in clay soils, anisotropy, vegetation, and hysteresis.

4.4   PREDICTION OF PORE PRESSURES

4.4.1   Non-swelling soils

As discussed in the previous section, the flow of water in unsaturated soils can be adequately described by the modified Darcy law:

$$v = -K(h)\frac{d\phi}{dl} \tag{4.24}$$

The law of continuity or conservation of mass also states that for an incompressible fluid:

$$\frac{d\theta}{dt} = -\frac{dv}{dl} \tag{4.25}$$

where θ = volumetric water content.

Therefore, in one dimension, equations 4.24 and 4.25 give (Klute (1952))

$$\frac{d\theta}{dt} = \frac{d}{dl}\left[K(h)\frac{d\phi}{dl}\right] \tag{4.26}$$

or in three dimensions:

$$\frac{d\theta}{dt} = \nabla[K(h).\nabla\phi] \tag{4.27}$$

∇ = mathematical operator

$= \partial/\partial x_i$ where x_i represents the three dimensions, $i = 1, 2,$ and 3.

However, equation 4.27 is not convenient in practice, as it requires two soil-water variables, i.e. θ and ϕ (or h), although only one flow parameter, $K(h)$ is necessary. Rewriting equation 4.27

$$\frac{\partial\theta}{\partial h}.\frac{dh}{dt} = \nabla[K(h).\nabla\phi]$$

which gives,

$$\frac{dh}{dt} = \frac{\partial h}{\partial\theta}.\nabla[K(h).\nabla\phi] \tag{4.28}$$

This is the three-dimensional diffusion equation, which can be conveniently and adequately used to analyse transient pore pressure distributions in non-swelling soils associated with engineering structures.

Equation 4.28 has been applied to analyse the consolidation behaviour of saturated clays (Rendulic (1936), Terzaghi (1943), Davis and Poulos (1965)). Equation 4.26 is also basically the one-dimensional consolidation equation which has widespread use in soil engineering.

The diffusion equation has also been widely used in agricultural and hydrological applications. Philip (1969) has reviewed the mathematical methods for solving the non-linear diffusion equations, which are available for many problems of practical and theoretical interest. However, in most engineering applications, the complex geometry, boundary conditions and material properties can only be properly handled by numerical methods (Richards (1968) and (1972a)).

4.4.2   Swelling soils

Three basic new elements must enter the extension of flow theory even to a one-dimension flow and volume change in swelling soils.

(1) The soil particles are in motion; therefore Darcy's law applies to flow relative to the soil particles (Gerseranar (1937)).
(2) The necessary and sufficient soil parameter comprise $\theta(h)$ and $e(h)$, where $e(h)$ is an independent functional characteristic of the soil.
(3) Effects of gravity, surface loading and internal stresses have a significant effect.

The mathematical methods developed for the non-swelling soils, either apply directly or provide a starting point.

The steps used in developing theory for swelling unsaturated soils are:

(a) Transient flow equation. Equation 4.24 now becomes (Zaslovsky (1964), Philip (1968), Smiles and Rosenthal (1968))

$$v_r = -K(h)\frac{d\phi}{dl} \qquad (4.29)$$

where v_r = flow of water relative to motion of soil particles.

(b) Continuity equation. Defining a new material co-ordinate system, m, such that

$$\frac{dm}{dl} = (1+e)^{-1} \qquad (4.30)$$

and a new moisture variable, $\mathscr{V}$, defined as the moisture ratio (Philip and Smiles (1969), Raats (1965)), where

$$\mathscr{V} = (1+e)\theta \qquad (4.31)$$

then
$$\frac{\partial \mathscr{V}}{\partial t} = -\frac{\partial v_r}{\partial m}$$

giving
$$\frac{\partial \mathscr{V}}{\partial t} = \frac{\partial}{\partial m}\left(\frac{K}{1+e}\frac{\partial \phi}{\partial m}\right) \qquad (4.32)$$

Philip and Smiles (1969) rewrote equation 4.32 in terms of the single variable, giving:

$$\frac{\partial \mathscr{V}}{\partial t} = \frac{\partial}{\partial m}\left[\frac{K\,d\phi}{(1+e)d\mathscr{V}}\frac{\partial \mathscr{V}}{\partial m}\right] \qquad (4.33)$$

Philip and Smiles (1969) have obtained closed form solutions for problems involving absorption (swelling) and desorption (consolidation) for a step function change of surface moisture potential in a three component, i.e. unsaturated initially uniform and unloaded effectively semi-infinite system. As they have suggested, equation 4.33 can be extended to loaded systems, but this has not been published as at 1972.

Several implications of their work are:

(1) The mass flow of water associated with particle movement can be of the same order as Darcy flow relative to particles.
(2) The behaviour of swelling soils is often contrary to that predicted for non-swelling soils.
(3) Contrary to present consolidation theories (Terzaghi (1923), Biot (1941)), K and $d\phi/d\theta$ are not constant and the mass flow due to particle movement is not insignificant even for small strains.

4.4.3 Hydrostatics in swelling soils

Philip (1969a) has considered the problem of overburden pressures and surface load for the simple hydrostatic equilibrium case. Buckingham (1907) and Philip (1969b) showed that for a one-dimensional vertical column of swelling soil fixed at its base the equation for the potential at equilibrium is:

$$\phi = u - Z + \Omega \qquad (4.34)$$

where Ω = overburden potential resulting from the work performed per unit weight of water added to realise soil movement against gravity and external load. Thus if e is assumed to be a function of η (not P)

$$\Omega = P(Z)\mathrm{d}e/\mathrm{d}\mathscr{V}$$

where $P(Z)$ = total vertical stress (as head). Now

$$P(Z) = P(0) + \int_0^Z \gamma \, \mathrm{d}Z \qquad (4.35)$$

where $P(0)$ = external load at $Z = 0$.

$$\gamma = (\mathscr{V} + \gamma_s)/(1 + e) \qquad (4.36)$$
$$= \text{apparent wet specific gravity}$$

$$\gamma_s = \text{particle specific gravity}$$

Therefore the condition for moisture equilibrium in the vertical column is (Philip (1969b)):

$$\phi = u - Z + \frac{\mathrm{d}e}{\mathrm{d}\mathscr{V}}\left[P(0) + \int_0^Z \gamma \, \mathrm{d}Z \right]$$
$$= \text{constant} = -Z \qquad (4.37)$$

where Z = water table depth.

Differentiating equation 4.37 for $\mathrm{d}Z/\mathrm{d}\mathscr{V}$, a general solution can be obtained. A singular solution has been shown to occur (Philip (1969b)) at a value of $\mathscr{V}$ for which the apparent wet specific gravity, γ is a maximum. This value is designated the pycnotatic point, $\mathscr{V}_p$. Three types of possible equilibrium moisture profiles in a swelling soil can be defined according to the value of $\mathscr{V}_o$ at the surface.

(a) Hydric

For $\mathscr{V}_o > \mathscr{V}_p$, then $\mathscr{V} > \mathscr{V}_p$ and $\mathrm{d}\mathscr{V}/\mathrm{d}Z < 0$ for $Z \geqslant 0$

(b) Pycnotatic

For $\mathscr{V}_o = \mathscr{V}_p$, then $\mathscr{V} = \mathscr{V}_p$ and $\mathrm{d}\mathscr{V}/\mathrm{d}Z = 0$ for $Z \geqslant 0$

(c) Xeric

For $\mathscr{V}_o < \mathscr{V}_p$, then $\mathscr{V} < \mathscr{V}_p$ and $\mathrm{d}\mathscr{V}/\mathrm{d}Z > 0$ for $Z \geqslant 0$

For all three types of profiles,

$$\lim_{Z \to \infty} \mathscr{V} = \mathscr{V}p$$

Physical consequences of the above assumptions for the hydrostatics in swelling soils include:

(1) Effect of gravity in swelling soil is $(1 - \gamma \, \mathrm{d}e/\mathrm{d}\mathscr{V})$ times that in non-swelling soil. For a normal numeral soil, this factor can vary from -1 to $+1$.

(2) Overburden effects contrary to previous theory manifest themselves right to the surface (depend on $d\Omega/dZ$ not Ω).

(3) Various classic concepts of ground water hydrology fail for swelling soils, e.g. the concept of water table.

4.4.4 Steady vertical flow in swelling soils

Combining equations 4.29 and 4.37 give,

$$v_r = -K\left[\left\{\frac{du}{d\mathscr{V}}+\frac{d^2e}{d\mathscr{V}}\left(P(0)+\int_0^Z \gamma\,dZ\right)\right\}\frac{\partial\mathscr{V}}{\partial Z}+\gamma-\frac{de}{d\mathscr{V}}-1\right] \qquad (4.38)$$

Equation 4.38 can be solved for $dZ/d\mathscr{V}$ which on integration gives $Z(\mathscr{V})$. A singular solution occurs (Philip (1969c)) when:

$$v = K(1-\gamma\,de/d\mathscr{V}) = K(\mathscr{V}) \qquad (4.39)$$

$K(\mathscr{V})$ has one and only one maximum which occurs at

$$\mathscr{V} = \mathscr{V}_k < \mathscr{V}_p \qquad (4.40)$$

As it follows that for any possible value of v:

$$v = K(\mathscr{V}_\infty) \qquad \text{where} \qquad \mathscr{V}_\infty = \lim_{Z\to\infty}\mathscr{V}$$

Therefore $\mathscr{V}_k$ is the value of $\mathscr{V}_\infty$ for which the greatest possible steady downward flow occurs. This point is defined as the katotatic point.

The classification of the possible solutions is rather elaborate. However, two important implications emerge:

(1) For $\mathscr{V}_\infty < \mathscr{V}_k$, steady flow can only occur when $\mathscr{V}_0 = \mathscr{V}_\infty$ and $d\mathscr{V}/dZ = 0$.

(2) For $\mathscr{V}_0 > \mathscr{V}_\infty > \mathscr{V}_p$, steady upward flow can occur against the gradient of $\mathscr{V}$.

4.4.5 Unsteady vertical flow in swelling soil

By combining equation 4.38 with the continuity requirement for dm, the general equation for unsteady vertical flow and volume change in swelling soils can be obtained:

$$\frac{\partial\mathscr{V}}{\partial t}=\frac{\partial}{\partial m}\left[\frac{K}{1+e}\left\{\frac{du}{d\mathscr{V}}+P(0)\frac{d^2e}{d\mathscr{V}^2}\right\}\frac{\partial v}{\partial m}\right]$$
$$-\frac{\partial}{\partial m}\left[K\left\{1-\gamma\frac{de}{d\mathscr{V}}-\frac{d^2e}{d\mathscr{V}^2(1+e)}\cdot\int_0^m (1+e)dm\frac{\partial\mathscr{V}}{\partial m}\right\}\right] \qquad (4.41)$$

Philip (1969c) has produced solutions for equation 4.41 for the case of infiltration. Two classes of asymptotic solutions, i.e. for large time, t, are:

(1) For $\mathscr{V}_\infty \geqslant \mathscr{V}_k$, the solution is a steady distribution according to equation 4.38.

(2) For $\mathscr{V}_\infty < \mathscr{V}_k$, the solution asymplotically approaches a steady distribution for $\mathscr{V}_k \leqslant \mathscr{V} < \mathscr{V}_0$ connected to a wetting front for $\mathscr{V}_\infty \leqslant \mathscr{V} < \mathscr{V}_k$ as for non-swelling soils.

4.4.6   Two- and three-dimensional flow in swelling unsaturated soils

The extension of the one-dimensional theory to two and three dimensions involves several new assumptions (Philip (1969b)) including a theory of stress distribution. This redistribution of stress following moisture flow and volume changes was also shown to be important for three-dimensional consolidation problems in saturated soils (Biot (1941), Christian (1965)).

Rewriting equation 4.32 in terms of the physical co-ordinate, l (Richards (1972))

$$\frac{\partial \mathscr{V}}{\partial t} = (1+e)\frac{\partial}{\partial l}\left(K\frac{\partial \phi}{\partial l}\right) \tag{4.42}$$

and
$$\frac{\partial \theta}{\partial t} = \frac{(1+e)}{(1+e+\theta\,de/d\theta)}\frac{\partial}{\partial l}\left(K\frac{\partial \phi}{\partial l}\right) \tag{4.43}$$

in one dimension, or:

$$\frac{\partial \theta}{\partial t} = \frac{(1+e)}{(1+e+\theta\,de/d\theta)}.\nabla(K.\nabla\phi) \tag{4.44}$$

in three dimensions, where

$$\phi = u - Z + \Omega$$

$$= u - Z + \frac{\sigma Z}{\gamma_w}\frac{de}{d\mathscr{V}}$$

and
$$\frac{\sigma_Z}{\gamma_w} = P(Z) \quad \text{in m-water head}$$

Equation 4.44 is therefore equivalent to equation 4.42 developed by Philip (1969c). As $de/d\theta$ can only take the value

$$0 \leqslant de/d\theta \leqslant 1$$

it can be seen that $d\theta/dt$ for a swelling soil is less than $d\theta/dt$ for a non-swelling soil. Use of the numerical solution for swelling soils (Richards (1972b)) as discussed later in Section 4.6, gave rates of consolidation and swelling for a swelling soil less than for a non-swelling soil, although the possible effects of the numerical procedure could not be isolated.

4.5   IMMEDIATE DEFORMATION UNDER LOAD

4.5.1   Linear analyses

For most practical cases, linear elastic models provide useful predictions of the deformation of engineering structures. While for most soils, the stress-strain behaviour is non-linear, good results can often be obtained from linear elastic parameters, i.e. the elastic modulus, E and Poisson's ratio, v, determined along the appropriate stress and moisture paths, as for example in the analysis of road pavements (Monismith et al. (1967); Richards and Gordon (1972)). This approach is applicable to both saturated and unsaturated soils.

Perhaps the most flexible method of solving linear elastic models is the finite element method, which has been adequately described elsewhere (Zienkiewicz and Cheung (1967), Clough and Rashid (1965) and Wilson (1965)). By representing the continuum as an assemblage of small finite elements, the equilibrium relationships between the displacements of the nodes of the elements and the applied forces can be written in matrix form:

$$[K].\{\delta\} = \{F\} \tag{4.45}$$

where $[K]$ = stiffness matrix assembled for the whole system;

$\{\delta\}$ = vector of all nodal displacements;

$\{F\}$ = vector of all nodal forces.

The stiffness matrix is a function of only two material parameters, E and v, for isotropic soils, and four parameters, E_x, E_z, v_x and v_z for a simple representation of cross-anisotropic soils. Knowing the prescribed loads or displacements at each boundary nodal point, equation 4.45 can be solved for both vertical and horizontal displacements at each nodal point. The stresses and strains can be readily computed from these displacements.

4.5.2 Linear material parameters

As mentioned above, linear isotropic materials can be represented by only two parameters, E and v, which can be conveniently determined in the laboratory using the conventional triaxial test procedure in which both vertical and lateral strains are monitored. For the analysis of existing structures, these parameters can be determined on sealed 'intact' samples taken from representative points within the loaded soil mass, using the appropriate triaxial stress state.

In determining the effective parameters of a future structure on unsaturated soils the problem is more difficult. Not only must the stress level, which may depend on the material parameters, be predicted, but also the soil suction and its history at each point. Various methods of predicting soil suction are available (Richards (1967), (1972a) and (1972b)) including those discussed in the previous and following sections. This predicted soil suction can be established in a triaxial specimen either sampled at its initial condition or compacted according to its placement conditions by placing in suction plate, pressure membrane or vacuum dessicator apparatus. In this way, the true moisture path and the stress path, if possible, can be simulated in the laboratory.

Figure 4.8 shows the effect of soil suction on the modulus of a triaxial specimen of a clay sampled at its initial condition. These results suggest that a good correlation can be established between the modulus and soil suction and expressed in the form:

$$E = K.h^n \tag{4.46}$$

where E = effective modulus;

K = coefficient;

h = soil suction;

n = exponent.

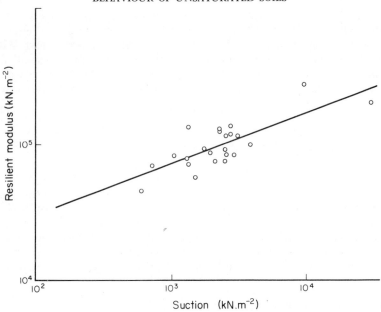

Figure 4.23   Resilient modulus versus soil suction for a clay. Type M

Similar correlations can also be obtained for the resilient modulus, MR, as determined in a repeated loading triaxial apparatus for road pavement analyses *(Figure 4.23)*.

4.5.3   Non-linear analyses

As mentioned earlier, most soils exhibit non-linear properties and should be analysed accordingly for best results.

An approximate non-linear analysis can be carried out by repetitive linear analyses, using either iterative, incremental type solutions or a combination of both (Dehlen and Monismith (1970), Richards (1972b)). In this method, new material properties are computed after each linear analysis.

For stress and moisture-dependent material parameters, it is necessary to use as data initial stresses and soil suction. The initial stresses can be measured values fed in as data or computed within the programme. Soil suction can also be measured or predicted as already discussed.

In addition to non-linear elastic analyses, no—or limited—tension, visco-elastic or elasto-plastic analyses can also be handled in the same way, as will be discussed in Chapter 5.

4.5.4   Non-linear material parameters

The testing required to characterise materials as non-linear is complex. To completely assess a non-linear elastic material, it would be necessary to measure all six strains induced by all possible combinations of six normal and shear stresses. However, some attempts have been made to simplify evaluation of the non-linear response of soils.

In general, the elastic moduli can be simply related to the stress invariants (Huang (1969)) or the octahedral stresses (Morgan and Gerrard (1969), Holden (1967), Morgan and Holden (1967)). In the latter case, the relation can be expressed in the form

$$E = a + b\sigma_{oct}^n - c\tau_{oct}^m \qquad (4.47)$$

where a, b and c are material constants;

   σ_{oct} = octahedral normal stress;

   τ_{oct} = octahedral shearing stress;

   n, m = exponents.

As already discussed above, the elastic moduli can also be related to soil suction according to equation 4.46. Equations 4.46 and 4.47 can therefore be combined (Richards (1972a) and (1972b)) to give the alternative relations for non-linear response:

$$E = (a + b\sigma_{oct}^n + c\tau_{oct}^m) \cdot h^p \qquad (4.48)$$

or $\qquad\qquad E = a + b\sigma_{oct}^n + c\tau_{oct}^m + dh^p \qquad (4.49)$

An appropriate relation between moduli and stress and soil suction obtained from simple triaxial data can generally be determined using one of these two relations.

No equivalent study of Poisson's ratio appears to have been made. However, preliminary measurements (Richards (1972a) and (1972b)) have indicated that Poisson's ratio for an unsaturated expansive clay may vary from near 0.5 for a clay a low density and soil suction to 0.16 for the sample at a higher density and soil suction.

A typical analysis of a concrete slab-on-ground foundation on an expansive clay with non-linear properties is shown in *Figure 4.24*.

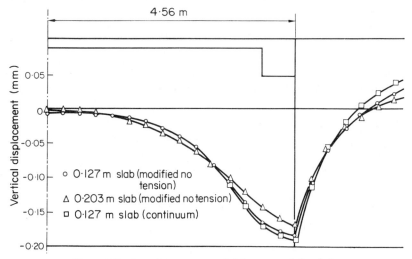

Figure 4.24   Immediate response of slab-on-ground foundation

4.6   TIME DEPENDENT DEFORMATIONS

4.6.1   One-dimensional consolidation theory

Although the consolidation theory for saturated soils was previously discussed in some detail (Lee (1968)), *(Figure 3)* the theory is again presented for comparison with the theory for unsaturated soils.

Terzaghi (1923) was the first to apply Darcy's law to the consolidation problem in saturated clays.

Combining the previous equations 4.41 and 4.42,

$$\frac{d\theta}{dt} = \frac{d}{dl} K \frac{d\phi}{dl}$$

for a saturated clay, or

$$\frac{d\theta}{dt} = K \frac{d^2\phi}{dl^2} \tag{4.50}$$

as K is a constant. Writing

$$\frac{d\theta}{dt} = \frac{d\theta}{du} . \frac{du}{dt}$$

equation 4.50 becomes

$$\frac{du}{dt} = \frac{du}{d\theta} . K \frac{d^2\phi}{dl^2} \tag{4.51}$$

or

$$\frac{du}{dt} = Cv \frac{d^2u}{dl^2} \tag{4.52}$$

neglecting gravity, i.e. $\phi = u$.

Equation 4.51 is the one-dimensional diffusion equation and equation 4.52, the one-dimensional consolidation equation. Soil engineers have tended to use equation 4.52 for the development of the consolidation theory for saturated soils; however, it is necessary to make the following assumptions:

(1) Soil is saturated.
(2) Both the pore fluid and soil particles are incompressible, but the soil skeleton is compressible.

(Note that these two assumptions make the saturated soil essentially incompressible to instantaneously applied loads.)

(3) Effective stress law is valid, i.e. $\sigma' = \sigma - u$.
(4) Strains are infinitesimal.
(5) The flow parameters are linear and reversible, i.e. a constant for the material.
(6) There is no spatial distribution of stress, e.g. to ensure equilibrium

$$\frac{d\sigma}{dx} = 0$$

(7) There is no redistribution of stress, e.g.

$$\frac{d\sigma}{dt} = 0$$

The solutions of equation 4.52 are now readily available for simple geometry and boundary conditions from solutions of heat-flow theory (Carslaw and Jaeger (1958)).

4.6.2 Three-dimensional diffusion theory

Most problems are the result of flow in two and often three dimensions. For these cases, it is possible (Rendulic (1936), Terzaghi (1943), Davis and Poulos (1965)) to extend equation 4.52 to three dimensions:

$$\frac{du}{dt} = C_{v_3}\nabla^2 u \tag{4.53}$$

where C_{v_3} is the coefficient of consolidation for three-dimensional strain, which is different from C_v for one-dimensional consolidation.

4.6.3 Biot's consolidation theory

Equation 4.53, however, does not include the redistribution of total stresses, which is an important factor in two- or three-dimensional consolidation, because the differential volume changes throughout the field of the problem cause changes in total stress.

Biot (1941) developed a more general theory for three dimensions, which includes this interaction between volume change and stresses.

The basic equations of Biot (Christian (1966)) are:

(a) Transient fluid flow equation (from equation 4.50)

$$\frac{d\varepsilon_v}{dt} = \frac{K}{\gamma_w}\Delta^2 u \tag{4.54}$$

where ε_v = volumetric strain;

γ_w = pore fluid density.

(b) The effective stress principle

$$\sigma' = \sigma - u \tag{4.55}$$

where σ' = mean effective stress;

σ = mean total stress.

(c) Stress-strain relation

$$\frac{d\varepsilon_v}{dt} = \frac{3(1-2v')}{E'}\frac{d\sigma'}{dt} \tag{4.56}$$

where E' and v' are the drained modulus and Poisson's ratio of the material.

(d) Equilibrium

$$\frac{\partial u}{\partial x_i} + \frac{E'}{2(1+v')}\nabla^2 S_i + \frac{E'}{2(1+v')(1-2v')}\frac{\partial \varepsilon_{ii}}{\partial x_i} = 0 \tag{4.57}$$

where $x_i = x$, y and z directions;

S_i = displacements;

ε_{ii} = normal strains.

(e) Consolidation equation

$$\frac{K}{\gamma_w} \frac{E'}{3(1-2v')} \cdot (\nabla^2 u) = \frac{du}{dt} - \frac{d\sigma}{dt} \tag{4.58}$$

where $\dfrac{K.E'}{3\gamma_w(1-2v')} = C_{v_3}$

$\dfrac{K.E'}{\gamma_w(1-2v')(1+v')} = C_{v_1}$

The above equations show that the pore pressure, u, and the deformations cannot be solved from just one diffusion equation as in Terzaghi's one-dimensional theory, where the equilibrium is automatically satisfied. It can be seen that the volumetric strain, ε_v is not stated explicitly in terms of du/dt, but also requires the term $d\sigma/dt$ in its simplest form. Its significance is as follows:

(1) $\partial\sigma/dt$ shows that a redistribution of total stress should be considered in order to satisfy equilibrium, e.g. the Mandel-Cryer effect.
(2) The solutions available for the regular diffusion equation 4.52 are not applicable.
(3) Only the simultaneous solution of the above equation 4.58 can provide a complete solution.

(a) Mandel–Cryer effect.   Mandel (1953) showed that at locations where the pore fluid cannot dissipate at the beginning, there is an increase in the mean total stress, σ. This effect is known as the Mandel-Cryer effect.

(b) Three-dimensional effects.   In general, the effects of three-dimensional boundary conditions and flow are to increase the rate of the diffusion process, but the redistribution of total stresses can reduce the rate where the Mandel-Cryer effect occurs. Therefore, each application must be considered separately. Furthermore, C_v is difficult to measure in more than one-dimensional cases as:

(1) C_v is dependent on the boundary displacement conditions, e.g. C_v is different for plane strain and stress and axisymmetric problems.
(2) Anisotropic stress-strain or flow behaviour of the soil cannot be expressed by single soil parameters.
(3) It is difficult to determine C_v in three-dimensional tests.

(c) Limitations of Biot's theory.   The two main limitations of Biot's consolidation theory are:

(1) It is restricted to linear or constant material properties.
(2) Boundary conditions other than a few simple conditions are very difficult to handle.

4.6.4 Finite element solution of Biot's theory

Due to its incremental character and the more convenient handling of boundary conditions, the finite element formulation (Christian and Boehmer (1970)) is applicable to non-linear, non-homogeneous material properties and can take into account a wider range of boundary conditions. However, it must be realised that it only solves the Biot equations in a more convenient manner.

4.6.5 Two- and three-dimensional consolidation theory for unsaturated soils

Using the concepts of Biot (1941), Christian and Boehmer (1969) and Philip (1969c), a new consolidation and swelling theory and method of solution has been developed (Richards (1972b)). The basic equations are as follows:

(a) Transient fluid flow equation

$$\frac{\partial \theta}{\partial t} = \frac{(1+e)}{(1+e+\theta\, de/d\theta)} \cdot \nabla(K(h)\nabla\phi) \tag{4.59}$$

(b) Soil water potential equation

$$\phi = u - z + \frac{\phi z}{\gamma_w}\, de/d\mathscr{V} \tag{4.60}$$

(c) Volume change equation

$$\frac{\partial \varepsilon_v}{\partial t} = \frac{de}{d\theta} \cdot \frac{\partial \theta}{\partial t} \tag{4.61}$$

(d) Stress-strain relation

$$\frac{\partial \sigma'}{\partial t} = \frac{E'}{3(1-v')} \frac{\partial \varepsilon_v}{\partial t} \tag{4.62}$$

(e) Equilibrium

$$\frac{\partial u}{\partial x_i} + \frac{E'}{2(1+v')} \nabla^2 S_i + \frac{E'}{2(1+v')(1-2v')} \cdot \frac{\partial \varepsilon_{ii}}{\partial x_i} = 0 \tag{4.63}$$

(f) Consolidation equation

$$\frac{E'(h)}{3[1-v'(h)]} \cdot \frac{de_v}{d\theta} \cdot \frac{(1+e)}{(1+e+\theta\, de/d\theta)} \nabla\left[K(h) \cdot \nabla\left(u-z+\frac{\sigma z}{\gamma_w}\frac{de}{d\mathscr{V}}\right)\right] = \frac{\partial \sigma'}{\partial t} \tag{4.64}$$

These six basic equations are formed by modifying the Biot equations for three-dimensional consolidation using the theories of Philip (1969c) and Philip and Smiles (1969) for the consolidation and swelling of unsaturated soils. Thus they provide the most general theory for flow and equilibrium of water and load-deformation behaviour of unsaturated swelling soils. These equations can be solved (Richards (1972b)), by a simple stepwise incremental solution.

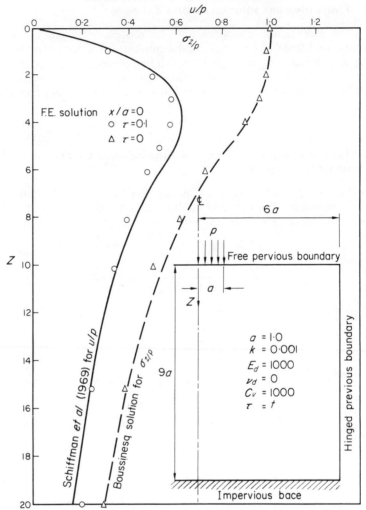

Figure 4.25   Initial changes in pore pressure for a strip load on a saturated semi-infinite media

(1) Soil water potential, including initial values, can be defined as:

$$\phi = -h - z + \frac{\sigma}{\gamma_w}\frac{de}{d\mathscr{V}} \tag{4.65}$$

$$= -h - z \quad \text{if } h \text{ measured } in \text{ } situ \text{ under stress}$$

An example of the initial changes in pore pressure for a strip load on saturated semi-infinite medium is shown in *Figure 4.25*.

(2) Transient fluid flow equation for swelling unsaturated soil is:

$$\frac{\partial \theta}{\partial t} = \frac{(1+e)}{\left(1+e+\theta\dfrac{de}{d\theta}\right)} . \nabla(K(h) . \nabla \phi) \tag{4.66}$$

However, if during the first small time step, we assume the material does not change volume (e.g. the soil is constrained), then $[de/d\theta]_v = 0$, equation 4.66 becomes

$$\partial\theta/\partial t = \nabla(K(h) . \nabla\phi) \tag{4.67}$$

Equation 4.66 is readily soluble by the finite difference (Richards (1968)) or finite element technique (Richards (1972a)) giving $\nabla\theta$ for the small time interval, Δt. (Details of the finite element technique are given in Chapter 5, Section 5.6.)

(3) As a result of the change in water content, $\Delta\theta$, the soil tends to change volume:

$$\frac{\partial\varepsilon_v}{\partial t} = \left[\frac{de}{d\theta}\right]_\sigma . \frac{\partial\theta}{\partial t}$$

or

$$\Delta\varepsilon_v = \left[\frac{de}{d\theta}\right]_\sigma . \Delta\theta \tag{4.68}$$

where $[de/d\theta]_\sigma$ is determined for constant stress, σ.

(4) However, as stated above, since no volume change is allowed, the stress increase, $\Delta\sigma$, required to prevent $\Delta\varepsilon_v$ is given by (assuming h still remains constant):

$$\Delta\sigma_v = \frac{E'}{3(1-v')}\Delta\varepsilon_v \tag{4.69}$$

where E' and v' are not the normal secant drained parameters but incremental values following the true and moisture paths.

(5) The Biot equilibrium equation can be solved with the distributed load $\Delta\sigma_v$ applied giving equilibrium stresses, strains and displacements throughout the system.

(6) During the above steps, h has been kept constant. It can now be corrected by the equation:

$$\Delta\phi = \Delta h = \frac{\Delta\sigma}{\gamma_w}\frac{de}{d\mathcal{V}} \tag{4.70}$$

which in effect establishes equation 4.65 on page 152.

(7) As the incremental strains $\Delta\varepsilon$ throughout the system have been calculated, then a new co-ordinate system can be established:

$$x_1 = x_o(1+\Delta\varepsilon_1)$$

$$\left(\text{Note } x_o = \frac{x_1}{(1+\Delta\varepsilon_1)} = \frac{x_z}{(1+\Delta\varepsilon_2)} = \text{constant}\right)$$

compares with:

$$m = \frac{x}{(1+e)} \qquad \text{(Philip and Smiles (1969))}$$

(8) The whole process is then repeated for the next time interval. The shorter the time interval, the more accurate will be the solution.

4.6.6 Results

This solution method has been checked against many existing closed form solutions of the original Biot equation and it has been found satisfactory, particularly for finite boundary problems (Richards (1972*b*)).

Its extension to unsaturated soils is rational and based on theoretical considerations. Results from practical applications are reasonable, but cannot be checked other than by numerous, well-instrumented field trials.

The main limitations to its application are the determination of the various soil parameters:

(1) $K(h)$ as a function of h.
(2) $[de/d\theta]_\sigma$ as a function of h. Note: However, $[d\varepsilon_i/d\theta]_\sigma$ as a function of h is more convenient where ε_x, ε_y, and ε_z are the strains in the three principle directions.
(3) $E'(h_1\sigma)$ and $v'(h_1\sigma)_1$ the incremental elastic parameters as a function of h and σ. (Note: E' and v' can be obtained from same test as $[d\varepsilon_i/d\theta]_\sigma$.)
(4) $(de/d\mathscr{V})(h)$ as a function of h. (Note: This can be compared with Skempton's A and B parameters.)

All these four parametric functions can be obtained from one properly designed laboratory test. But due to scale effects, field testing is preferable, should it become possible in the future.

An example of typical solutions for a slab-on-ground foundation on expansive clay in a semi-arid environment six months following construction are shown in *Figure 4.26*. These solutions permit the examination of the various factors contributing to the behaviour of the foundation.

Of particular interest in *Figure 4.26* is the effect the stiffness of the slab has on the deformation. While the modified no-tension analyses gave deforma-

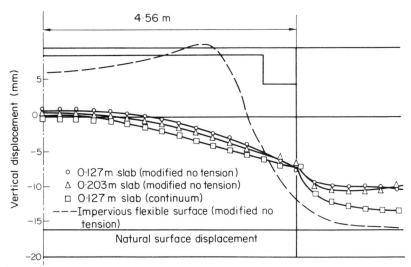

Figure 4.26 Deformations of a slab-on-ground foundation in a semi-arid environment six months after construction

tions little different from the non-linear continuum analyses, the stresses in the soil and the slab were far more realistic in relation to the tensile strength of the materials.

BIBLIOGRAPHY

AITCHISON, G. D. (1956). 'The circumstances of unsaturation in soils with particular reference to the Australian environment', *Proc. 2nd Aust.–N.Z. Conf. Soil Mech. and Fndn. Engng.*, 173–191.

AITCHISON, G. D., and RICHARDS B. G. (1965). 'A broad scale study of moisture conditions in pavement subgrades throughout Australia', *Moisture Equilibria and Moisture Change in Soils Beneath Covered Areas*, Butterworths, Australia, 184–282.

AITCHISON, G. D., RUSSAM, K., and RICHARDS, B. G. (1965). 'Engineering concepts of moisture equilibria and moisture changes in soils', *Moisture Equilibria and Moisture Changes in Soils Beneath Covered Areas*, Butterworths, Australia, 7.

AITCHISON, G. D., and RICHARDS, B. G. (1969). 'The fundamental mechanisms involved in heave and soil moisture movement and the engineering properties of soils which are important in such movement', *Conf. on Expansive Clay*, Texas A. and M. University, Texas.

BIOT, M. A. (1941). 'General theory of three-dimensional consolidation', *J. Appl. Phys.*, **12**, 155–164.

BOLT, G. H., and LAGERWERFF, J. V. (1964). 'Consequences of electrolyte redistribution during pressure membrane equilibration of clays', *Soil Sci.*, **99** No. 3, 147.

BUCKINGHAM, E. (1907). 'Studies on the movement of soil moisture', *U.S. Dept. Agr. Bur. Soil Bull.*, 38.

CARSLAW, H. S., and JAEGER, J. C. (1958), *Conduction of heat in solids*, Oxford Press.

CHILDS, E. C. J., and COLLIS-GEORGE, N. C. (1950). 'The permeability of porous materials', *Proc. Ray. Soc. A201*, 392–405.

CHRISTIAN, J. T. (1965). 'Two-dimensional analysis of stress and strain in soils, Rept. No. 1, Iteration procedure for saturated elastic porous materials', *MIT Dept. of Civ. Engng. Report No. R65–66.*

CHRISTIAN, J. T. (1967). 'Numerical methods of calculating time-dependent settlement and heave of embankments', *Internal Report, MIT, DSR76102.*

CHRISTIAN, J. T., and BOEHMER, J. W. (1970). 'Plane strain consolidation by finite elements', *J. Soil Mech. and Fdns. Div., Proc. ASCE*, **96**, SM 4, 1435–1457.

CLOUGH, R. W., and RASHID, Y. (1965). 'Finite element analysis of axisymmetric solids', *J. Engng. Mech. Div., ASCE*, 71–85.

COLEMAN, J. D., and MARSH, S. D. (1961). 'An investigation of the pressure membrane method for measuring the suction properties of soil', *J. Soil Sci.*, **12** No. 2, 343–362.

CRONEY, D., COLEMAN, J. D., and BRIDGE, P. M. (1952). 'The suction of moisture held in soil and other porous materials', *Tech. Pap. No. 24., Rd. Res. Lab., DSIR, England.*

DAVIS, E. H., and POULOS, H. G. (1965). 'Triaxial testing and three-dimensional settlement analysis', *Proc. 4th Aust.–N.Z. Conf. Soil Mech. and Fndn. Engng*, 233–243.

DEHLEN, G. L., and MONISMITH, C. L. (1970). 'The effect of non-linear material response on the behaviour pavements under traffic', *Proc. 49th Annual Meeting, HRB.*

GARDNER, W. R. (1956). 'Calculation of capillary conductivity from pressure plate outflow data', *Proc. Soil Sci. Soc. Amer.*, **20**, 317–320.

GERSEVANOR, N. M. (1937). *The foundations of Dynamics of Soils*, 3rd edition, Stroiizdat, Moscow-Leningrad, 1937.

GURR, C. G. (1962). 'Use of gamma rays in measuring water content and permeability in unsaturated columns of soil', *Soil Sci.*, **94** No. 4, 224–229.

HAINES, W. B. (1927). 'A further contribution to the theory of capillary phenomena in soils., *J. Agric. Sci.*, **17**, 264–290.

HOLDEN, J. C. (1967). 'Stresses and Strains in a sand mass subjected to a uniform circular load', *University of Melbourne, Civil Engng. Dept. Departmental Report No. 13.*

HOLMES, J. W. (1955). 'Water sorption and swelling of clay blocks', *J. Soil Sci.*, **6** No. 2, 200–208.

HUANG, Y. H. (1969). 'Finite element analysis of non-linear soil media', *Proc. ASCE Symp. on Application of Finite Element Methods in Civil Engng.*, Vanderbilt University.

KLUTE, A. (1952). 'A numerical method for solving the flow equation for water in unsaturated materials', *Soil Sci.*, **73**, 105–116.

KUNZE, R. J., and KIRKHAM, D. (1962). 'Simplified accounting for membrane impedance in capillary conductivity determinations', *Proc. Soil Sci. Soc. Amer.*, **26** No. 5, 421–426.

LANG, A. R. G. (1968). 'Psychrometric measurement of soil water potential *in situ* under cotton plants', *Soil Sci.*, **106**, 460–464.

MANDEL, J. (1953). 'Consolidation des Sols (etude mathematique)', *Geotechnique*, **111**, 287–299.

MARSHAL, T. J. (1958).

MILLER, E. E., and ELRICK, D. E. (1958). 'Dynamic determination of capillary conductivity extended for non-negligible membrane impedance', *Proc. Soil Sci. Soc. Amer.*, **22** No. 6, 383–386.

MONISMITH, C. L., SEED, H. B., MITRY, F. G., and CHAN, C. K. (1967). 'Prediction of pavement deflections from laboratory tests' *Proc. 2nd Intl. Conf. on Structural Design of Asphalt Pavements*, University of Michigan.

MORGAN, J. R., and HOLDEN, J. C. (1967). 'Deflection prediction in prototype pavements', *Proc. 2nd Intl. Conf. Structural Design of Asphalt Pavements*, University of Michigan.

MORGAN, J. R., and GERRARD, C. M. (1969). 'Stresses and Displacements in a rock mass having stress dependent properties', *Joint Symposium on Rock Mechanics, I.E. Aust. and AIMM*, Sydney, 87.

PECK, A. J., and RABBAGE, R. M. (1966). 'Soil water potential: direct measurement by a new technique', *Science*, **151**, 1385.

PECK, A. J. (1968). 'Theory of the Spanner psychrometer: I. The thermocouple,' *Agricultural Meteorology*, **5** No. 6, 433–447.

PHENE, C. J., HOFFMAN, G. J., and RAWLINS, S. L. (1971). 'Measuring soil matric potential *in situ* by sensing heat dissipation within a porous body: I. Theory and sensor construction,' *Soil Sci. Soc. Amer. Proc.*, **35**, 27–33.

PHILIP, J. R. (1958). 'Physics of water movement in porous solids', *Highway Res. Bd. Special Rept.*

PHILIP, J. R. (1968). 'Kinetics of sorption and volume change in clay colloid pastes', *Aust. J. Soil Res.*, **7**, 99–120.

PHILIP, J. R., and SMILES, D. E. (1969). 'Kinetics of sorption and volume change in three-component systems', *Aust. J. Soil Res.*, **7**, 1–19.

PHILIP, J. R. (1969a). 'Theory of infiltration', *Adv. Hydro-sci.*, **5**, 215–296.

PHILIP, J. R. (1969b). 'Moisture equilibrium in the vertical in swelling soils: I. Basic theory and II. Applications', *Aust. J. Soil Res.*, **7**, 99–141.

PHILIP, J. R. (1969c). 'Hydrostatics and hydrodynamics in swelling soils', *Water Resources Res.*, **5** No. 5, 1070–1077.

RAATS, P. A. C. (1965). 'Development of equations describing transport of mass and momentum in porous media, with particular reference to soils', *Ph.D. Thesis*, University of Illinois, Urbana.

RAWLINS, S. L., and DALTON, F. N. (1967). 'Psychrometric measurement of soil water potential without precise temperature control', *Soil Sci. Soc. Amer. Proc.*, **31**, 297–301.

RENDULIC, L. (1936). 'Parenziffer and Poren Wasserdruck in Tonen', *Der Bavingenieur*, **17**, 559–564.

RICHARDS, B. G. (1965). 'Measurement of the free energy of soil moisture by the psychrometric technique using thermistors', in *Moisture Equilibria and Moisture Changes Beneath Covered Areas*, Butterworths, Australia, 39–46.

RICHARDS, B. G. (1965). 'Determination of the unsaturated permeability and diffusivity functions from pressure plate outflow data with non-negligible membrane impedance', in *Moisture Equilibria and Moisture Changes in Soils Beneath Covered Areas*, Butterworths, Australia, 47–54.

RICHARDS, B. G. (1967). 'Moisture flow and equilibria in unsaturated soils for shallow foundations', in *Permeability and Capillarity of soils*, ASTM, STP417, 4–34.

RICHARDS, B. G. (1968). 'A mathematical model for moisture flow in Horsham clay', *Inst. Engnrs. Aust., Civil Engng. Trans. CE10*, **2**, 220–224.

RICHARDS, B. G. (1968). 'The solution of the two-dimensional moisture flow equation applied to road subgrade soil conditions', *Inst. Engnrs. Aust., Civil Engng. Trans. CE10*, **2**, 209–212.

RICHARDS, B. G. (1968). 'Review of measurement of soil water variables and flow parameters', *Proc. 4th Conf. ARRB*, **4** No. 2, 1843–1861.

RICHARDS, B. G. (1971). 'Psychrometric techniques for field measurement of negative pore pressures in soils', *Proc. 1st Aust.–N.Z. Geomechanics Conf.*

RICHARDS, B. G., and GORDON, R. (1972). 'Prediction and observation of the performance of a flexible pavement on an expansive clay subgrade', *Proc. 3rd Intl. Conf. on Structural Design of Asphalt Pavements*, London.

RICHARDS, B. G. (1972a). 'The analysis of flexible road pavements in the Australian environment:

Changes of pore pressure or soil suction', *Div. of Appl. Geomechanics, Tech. Paper No. 17*, CSIRO.

RICHARDS, B. G. (1972b). 'Transient behaviour of saturated and unsaturated soils under changing load and moisture conditions', *Div. of Appl. Geomechanics, Tech. Paper No. 16*, CSIRO.

RICHARDS, L. A. (1931). 'Capillary conduction of liquids through porous systems', *Physics 1*, 318–333.

RICHARDS, L. A., and OGATA, G. (1958). 'A thermocouple for vapour pressure measurement in biological and soil systems at high humidity', *Science*, **128**, 1089.

RICHARDS, L. A., and RICHARDS, P. L. (1962). 'Radial flow cell for soil water measurement', *Proc. Soil Sci. Soc. Amer.*, **26** No. 6, 515–518.

RIJTEMA, P. E. (1959). 'Calculation of capillary conductivity from pressure plate outflow data with non-negligible membrane impedance', *Neth. J. Agr. Sci.*, **7**, 209–215.

SMILES, D. E., and ROSENTHAL, M. J. (1968). 'The movement of water in swelling materials', *Aust. J. Soil Res.*, **3**, 1–9.

SPANNER, D. C. (1951). 'The Pettier effect and its use in the measurement of suction pressure', *J. Exp. Bot.*, **2**, 145.

TERZAGHI, K. (1923). 'Die Berechnung der Durchlässigkeit—sziffer des Tones aus dem Verlaw der hydrodynamische Spannungsercheinunge', reprinted in *From Theory to Practice in Soil Mechanics*, J. Wiley & Sons, New York, 133–146.

TERZAGHI, K. (1943). *Theoretical soil mechanics*, J. Wiley & Sons, New York.

WILSON, E. L. (1965). 'Structural analysis of exisymmetric solids', *AIAA Journal*, **3** No. 12, 2269–2273.

ZASLAVSKY, D. (1964). 'Saturated and unsaturated flow equation in an unstable porous medium', *Publ. No. 37*, Faculty of Civil Engng., Technion, Haifa, Israel.

ZIENKIEWICZ, O. C., and CHEUNG, Y. K. (1967). *The finite element method instructural and continuum mechanics*, McGraw-Hill, London.

Chapter 5

Analyses of Soil Settlement

I. K. Lee and S. Valliappan

5.1 INTRODUCTION

When reviewing the recent developments in total settlement and rate of settlement analyses, it is evident that the finite element technique is providing the opportunity for the solution of otherwise intractable problems. There have also been significant developments in the formal analyses of three dimensional rate problems, and some use has been made of finite difference techniques to predict the behaviour of layered soils.

The present discussion follows the earlier review of the theories describing the rate of consolidation (Lee (1968)). Interest has been shown in recent times in the constant rate of strain type of one-dimensional consolidation test. The theory and application of this test is discussed. To complement the earlier review on layered soils attention is directed to the numerical evaluation of the consolidation theory for a one-dimensional strain state. It is seen that the results for a two-layered soil deposit can be adequately presented. The Rowe-Horne theory for consolidation of a layered deposit in which sand drains are placed is then discussed and the important numerical results presented.

Formal solutions of the Biot theory applied to a range of two- and three-dimensional problems are available. Some of these solutions were reviewed previously. A more complete summary is now presented including the finite element theory as applied to this class of problems.

The finite element analysis is used to analyse the behaviour of a raft foundation supported on a soil represented by either a Winkler or linear elastic model. This analysis is then extended to take account of the stiffness of the superstructure in order to establish a logical consistent analysis for the settlement behaviour of the raft.

The finite element technique has allowed non-linear settlement analyses to be developed. As an approximation one stage better than the linear analyses one can consider the stress-strain characteristic to be represented by bi-linear relationship. The techniques for solving this and other non-linear situations are discussed, and the technique is applied to the problem of predicting the total settlement of a strip footing.

5.2 ONE DIMENSIONAL CONSOLIDATION

A generalised solution of the linear and non-linear consolidation equations was presented in 'Soil Mechanics; Selected Topics' (Butterworth) for a saturated soil subjected to small strains (Lee (1968) (Ch. 3)). The corresponding analysis for an unsaturated soil is discussed in Chapter 3 of the present volume, and the large strain case has been analysed (Gibson et al. (1967a)).

In 'Soil Mechanics; Selected Topics' it was recorded (p. 135) that there are certain advantages in conducting a consolidation test at a constant loading rate, and Schiffman's (1958) solution of the conventional diffusion equation for a constant loading rate was presented. More recently, Aboshi et al. (1970) have carried out a detailed evaluation of the solution. They transformed Schiffman's solution (p. 144) to the form

$$\frac{u}{p} = \frac{16}{\pi^3 T} \sum_{m=0}^{\infty} \frac{1}{(2m+1)^3} \left(1 - \exp\left(-\frac{(2m+1)^2 \pi^2 T)}{4}\right)\right) \sin\frac{(2m+1)y}{2H_D} \quad (5.1)$$

where u = excess pore pressure;

T = time factor = ct/H_D^2;

m = integer;

H_D = drainage path;

y = distance from element to drainage boundary;

p = external consolidation pressure.

The excess pore pressure at the base of an odometer (u_d) with drainage at the surface and an impermeable base is

$$\frac{u_d}{p} = \frac{16}{\pi^3 T} \sum_{m=0}^{\infty} (-1)^m \frac{1}{(2m+1)^3} \left(1 - \exp\left(\frac{(2m+1)^2 \pi^2 T)}{4}\right)\right) \quad (5.2)$$

Thus if p is recorded at a specific time the time factor can be determined provided u_d is measured.

The average degree of consolidation relative to the total stress applied at a specific time is defined as

$$U_c = 1 - \frac{\bar{u}}{p}$$

$$= 1 - \frac{32}{\pi^4 T} \sum_{m=0}^{\infty} \frac{1}{(2m+1)^4} \left(1 - \exp\left(-\frac{(2m+1)^2 \pi^2 T)}{4}\right)\right) \quad (5.3)$$

As an alternative procedure we could subject the soil to constant rate of strain (Smith and Wahls (1969), Wizza et al. (1971)). Assuming that the coefficient of consolidation, the permeability, and the coefficient of volume decrease are independent of the distance from the upper permeable boundary, the latter authors established that the vertical strain, at a point and at a specific time is,

$$\varepsilon = rt\left(1 + F\left(\frac{y}{H}, T\right)\right) \quad (5.4)$$

where r = rate of vertical strain, $H = H_D$ = thickness, and

$$F = \frac{1}{6T}\left(2 - 6\frac{y}{H} + 3\left(\frac{y}{H}\right)^2\right) - \frac{2}{\pi^2 T}\sum_{n=1}^{\infty}\frac{\cos\frac{n\pi y}{H}}{n^2}\exp(-n^2\pi^2 T) \quad (5.5)$$

showing that the strain at a point is equal to the average external strain at time t less a component which is a function of position and time. It will be noted, however, that the first expression of equation is not time-dependent, and expresses the gradient required to maintain a constant flow of water. Evaluation of F establishes that the time-dependent component becomes insignificant for values of T in excess of, say, 0.5. Thus, for these larger times,

$$\varepsilon \simeq \frac{rH^2}{c}\left(T + \frac{1}{6}\left(3\left(\frac{y^2}{H}\right) - 6\frac{y}{H} + 2\right)\right) \quad (5.6)$$

In order to determine the coefficient of consolidation at early time, that is, when the transient strain component is significant, Wizza et al. consider the ratio of the strain at the impermeable base of the sample to the strain at the permeable top of the sample, i.e.

$$\frac{\varepsilon(0, t)}{\varepsilon(H, t)} = \frac{T - \frac{1}{6} - \frac{2}{\pi^2}\sum\frac{\cos n\pi}{n^2}\exp(-n^2\pi^2 T)}{T + \frac{1}{3} - \frac{2}{\pi^2}\sum\frac{1}{n^2}\exp(-n^2\pi^2 T)} = F_3(T) \quad (5.7)$$

For a linear material

$$F_3 = \frac{(\sigma_{vt} - u_d) - \sigma_{vO}}{\sigma_{vt} - \sigma_{vO}} \quad (5.8)$$

where σ_{vt} is the total stress at time t and σ_{vO} is the initial vertical consolidation stress.

For a non-linear material

$$F_3 = \frac{\log_{10}(\sigma_{vt} - u_d) - \log_{10}\sigma_{vO}}{\log_{10}\sigma_{vt} - \log_{10}\sigma_{vO}} \quad (5.9)$$

Thus if the pore pressure u_d is measured, the value of F_3 is established from equation 5.8, hence T from a graphical representation of equation 5.7 and hence the coefficient of consolidation.

When the transients are effectively zero (equation 5.6) the value of the coefficient of consolidation for a linear soil is

$$c = \frac{H^2}{2u_d}\left(\frac{\Delta\sigma_{vt}}{\Delta t}\right) \quad (5.10)$$

which is equivalent to Smith and Wahl's expression.

For a non-linear soil the coefficient of consolidation is determined from a knowledge of the total stress at time t_O and at time $(t_O + \Delta t)$ and the pore pressure at the impermeable base, i.e.

$$c = -\frac{H^2\log_{10}\left(\frac{\sigma_{vO} + \Delta\sigma_v}{\sigma_{vO}}\right)}{2\Delta t\log_{10}\left(1 - \frac{u_d}{\sigma_v}\right)} \quad (5.11)$$

the values of pore pressure and total vertical stress in the denominator being average values over the time interval.

An oedometer designed to apply either a constant loading rate or a constant strain rate or a constant load is described by Wizza *et al.*

Mention should also be made of the special one-dimensional consolidation test devised by Lowe *et al.* (1969) in which a constant hydraulic gradient is applied.

5.3   CONSOLIDATION OF LAYERED DEPOSITS

5.3.1   One-dimensional consolidation

The solution for the consolidation of a multi-layered soil deposit establishes that two dimensionless parameters are required to define the inter-relationship between the consolidation behaviour of two adjacent layers i, and $(i+1)$. (Lee (1968), p. 144), i.e.

$$a_{i,(i+1)} = \frac{m_i}{m_{i+1}} \cdot \frac{h_i}{h_{i+1}}$$

and

$$B_{i,(i+1)} = \frac{k_{i+1}}{k_i} \cdot \frac{h_i}{h_{i+1}}$$

It will be recalled that m, h and k are the coefficient of volume decrease, the thickness and the vertical permeability of the particular layer. The ratio a/B will, for convenience, be designated as α.

For the two-layer deposit and the Rowe-Horne model of a layered deposit consisting of identical layer pairs, the 'a' and 'B' parameters have only one value each. For the general case of a soil deposit composed of n different layers, $(n-1)$ 'a' parameters and $(n-1)$ 'B' parameters will define the rate of settlement.

When selecting the definition of the time factor for such a deposit it is logical to choose the total thickness of the deposit, H, and the equivalent coefficient of consolidation of the whole deposit, $\bar{c}$, in preference to the values of any particular layer. Thus the time factor, $\bar{T}$, is defined as,

$$\bar{T} = \frac{\bar{c}t}{H^2} \tag{5.12}$$

where,

$$\bar{c} = \frac{H^2}{\left(\sum_1^n m_i h_i\right)\left(\sum_1^n \frac{h_i}{k_i}\right)}$$

For a two-layer deposit consisting only of layer 1 overlying layer 2, or a deposit having these layers alternating,

$$\bar{T} = \frac{T_1 aB}{(1+a)(1+B)} \tag{5.13}$$

where

$$T_1 = \frac{c_1 t}{p^2 h_1^2}$$

and $p = $ number of pairs.

In 'Soil Mechanics; Selected Topics', a finite difference expression for flow continuity was derived, but it has been established that use of this expression (equation 3.190b, p. 146) may lead to an instability in the numerical solution for values of 'a' in excess of 0.1. The expression recommended also satisfies continuity but does not lead to instability, i.e.

$$u_{rs} = \frac{1}{3(1+B)} \left(4(Bu_{r,s+1}+u_{r,s-1}) - (Bu_{r,s+2}+u_{r,s-2}) \right) \tag{5.14}$$

where r is an integer defining time, and s is an integer defining position within a layer. It should be noted that, in the previous notation, $u_{o,s}$ was used instead of $u_{r,s}$, u_1 instead of $u_{r+1,s}$ etc.

The degree of settlement is given by the expression,

$$U_s = 1 - \frac{\left(\frac{1}{2qu_o} \left(s(1) + \frac{1}{a_{1,2}} s(2) + \frac{1}{a_{1,2} \cdot a_{2,3}} s(3) + \ldots \right) \right)}{\left(1 + \frac{1}{a_{1,2}} + \frac{1}{a_{1,2} \cdot a_{2,3}} + \ldots \right)} \tag{5.15}$$

where u_o is the initial uniform pore pressure $s(1), s(2), \ldots, s(i)$ are defined as

$$s(i) = u_{s'} + 2u_{s'+1} + \ldots + u_{s''}$$

where s' is an integer defining the upper node of the ith layer $(=(i-1)(q+1))$, and s'' is an integer defining the lowest node of the ith layer $(=(i)(q+1))$. q defines the node spacing, i.e.

$$q = \frac{h_i}{\delta h_i} = \frac{h_{i+1}}{\delta h_{i+1}}$$

that is, an equal number of nodes per layer.

For a two-layer soil the expression simplifies to

$$U_s = 1 - \frac{1}{2u_o q(1+a)} (as(1)+s(2)) \tag{5.16}$$

There are special cases in which the settlement time relationship can be derived from the solution for a homogeneous layer.

It was first suggested by Glick (1945) in the discussion to Gray's paper (Gray (1945)) that if the depth, y, within the ith layer is transformed by the relationship

$$Y_T = y \sqrt{\left(\frac{\bar{c}}{c_i} \right)} \tag{5.17}$$

then the diffusion equation becomes common to all layers, that is, the transformed single layer of total thickness,

$$H_T = \sum_{i=1}^{i=n} h_i \sqrt{\left(\frac{\bar{c}}{c_i} \right)} \tag{5.18}$$

is equivalent to the layered system with respect to the basic diffusion equation. For complete equivalence the inter-layer continuity condition must also be satisfied. This leads to the condition that

$$\frac{k_i}{k_{i+1}} = \sqrt{\left(\frac{c_i}{c_{i+1}}\right)} \tag{5.19}$$

or

$$a_{i,i+1} = B_{i,i+1}$$

or

$$\alpha_{i,i+1} = 1 \quad \text{for all } i$$

It can then be shown that the total transformed thickness is equal to the actual total thickness

$$H_T = H$$

and thus the degree of settlement for this special case is given directly by the Terzaghi solution provided $\bar{c}$ is used as the coefficient of consolidation.

A second important special case occurs when the product $a.B$ is unity. Employing the thickness transformation of the above special case to a two-layer deposit subject to two-way drainage, the inter-layer continuity condition is satisfied irrespective of the value of α when the transformed thicknesses of the two layers are equal. This is satisfied by the condition $aB = 1$.

It then follows that the degree of settlement is given by the Terzaghi solution for a single layer of thickness H_T and characterised by a coefficient of consolidation of $\bar{c}$. For example, the time for 50% settlement is given by,

$$\frac{\bar{c}t_{50}}{H_T^2} = 0.0494$$

and

$$\frac{H_T^2}{H^2} = \frac{4a}{(1+a)^2}$$

Further special cases for the two-layer system are, using notation, P permeable, I impermeable,

$$PTIB \quad \text{(i)} \quad m_2 \to 0, \; a \to \infty, \; \alpha \to \infty, \quad B \text{ finite}$$
$$PTIB \quad \text{(ii)} \quad m_1 \to 0, \; a \to 0, \; \alpha \to 0, \quad B \text{ finite}$$
$$PTIB \quad \text{(iii)} \quad k_1 \to \infty, \; a \to \infty, \quad \text{but finite}$$
$$PTIB \quad \text{(iv)} \quad k_2 \to \infty, \; \alpha \to 0, \quad a \text{ finite}$$
$$PTPB \quad \text{(i)} \quad m_2 \to 0, \; a \to \infty, \; \alpha \to \infty, \quad B \text{ finite}$$
$$PTPB \quad \text{(ii)} \quad k_2 \to \infty, \; \alpha \to 0, \quad a \text{ finite}$$

Permeable Top, Impermeable Bottom ($PTIB$) (i).

$$m_2 = 0$$

Hence
$$a = \frac{m_1 h_1}{m_2 h_2} = \infty \quad \text{and} \quad \alpha = \frac{m_1 k_1}{m_2 k_2} = \infty$$

but
$$B = \frac{h_1 k_2}{h_2 k_1} \quad \text{is finite}$$

i.e. layer 2 has a finite permeability and thickness but is incompressible. Therefore, because the bottom boundary of layer 2 is impermeable, it effectively supplies an impermeable boundary to the bottom of layer 1. The consolidation of layer 1 must then be governed by the ordinary Terzaghi theory for one-way drainage.

Thus
$$U_s = U_T$$

where U_T is given by the Terzaghi theory for a time factor,

$$T_1 = \frac{c_1 t}{h_1^2} \tag{5.20}$$

and
$$\bar{T} = \frac{B}{1+B} T_1 \qquad \text{where} \qquad \bar{T} = \frac{\bar{c} t}{H^2}$$

For example, for $U_s = 0.5$,
$$\bar{T}_{50} = \frac{0.197 B}{1+B}$$

Permeable Top, Impermeable Bottom (*PTIB*) (ii).

$$m_1 = 0$$

Hence
$$a = \alpha = 0 \qquad \text{but } B \text{ is finite}$$

i.e. layer 1 has a finite permeability and thickness but is incompressible. It therefore impedes the consolidation of layer 2 but does not contribute to the total settlement. With boundary conditions

$$\left(\frac{\partial u}{\partial y}\right)_{y=0} = 0, \qquad \left(\frac{\partial u}{\partial y}\right)_{y=1} = -\frac{1}{B}(u)_{y=1} \qquad \text{and} \qquad (u)_{T=0} = 1$$

solution is

$$u = 2 \sum_{n \neq 1}^{\infty} \frac{\sin \lambda_n \cos(\lambda_n y)}{(\lambda_n + \sin \lambda_n \cos \lambda_n)} \exp(-\lambda_n^2 (1+B)\bar{T}) \tag{5.21}$$

and
$$U_s = 1 - 2 \sum_{n=1}^{\infty} \frac{\sin^2 \lambda_n}{\lambda_n (\lambda_n + \sin \lambda_n \cos \lambda_n)} \exp(-\lambda_n^2 (1+B)\bar{T}) \tag{5.22}$$

where the λ_n's are the roots of $\cot \lambda_n = \lambda_n \cdot B$.

Permeable Top, Impermeable Bottom (*PTIB*) (iii).

$$k_1 = \infty$$

Hence
$$\alpha = \infty \qquad \text{and} \qquad B = 0 \qquad \text{but } a \text{ is finite}$$

i.e. layer 1 is compressible but fully permeable so that it does not impede the drainage of layer 2 but makes an immediate contribution to the total settlement. Thus consolidation of layer 2 is governed by the ordinary Terzaghi theory for one-way drainage.

For the whole deposit,
$$U_s = \frac{a + U_T}{1+a} \tag{5.23}$$

where U_T is given by the Terzaghi theory for a time factor defined as

$$\frac{c_2 t}{h_2^2} = (1+a)\bar{T} \tag{5.24}$$

Permeable Top, Impermeable Bottom (*PTIB*) (iv).

$$k_2 = \infty$$

Hence $\alpha = 0$ and $B = \infty$ but a is finite

i.e. layer 2 is compressible but fully permeable so that at all depths it has a pore pressure equal to that in layer 1 at the interface, and the rate of flow of water out of layer 2 into layer 1 is proportional to the rate of compression of layer 2.

In layer 1

$$\frac{\partial u}{\partial \overline{T}} = \frac{(1+a)}{a} \frac{\partial^2 u}{\partial y^2}$$

where depth below top $= yh_1$. With boundary conditions

$$(u)_{y=0} = 0, \qquad \alpha\left(\frac{\partial u}{\partial y}\right)_{y=1} = \left(\frac{\partial^2 u}{\partial y^2}\right)_{y=1} \qquad \text{and} \quad (u)_{\overline{T}=0} = 1$$

solution is

$$u = 2 \sum_{n=1}^{\infty} \frac{(1-\cos \lambda_n)\sin(\lambda_n y)}{(\lambda_n - \sin \lambda_n \cos \lambda_n)} \exp\left(\frac{-\lambda_n^2(1+a)\overline{T}}{a}\right) \tag{5.25}$$

and

$$U_s = 1 - \frac{2}{1+a} \sum_{n=1}^{\infty} \frac{(1-\cos \lambda_n)(a - a \cos \lambda_n + \lambda_n \sin \lambda_n)}{\lambda_n(\lambda_n - \sin \lambda_n \cos \lambda_n)}$$

$$\times \exp\left(\frac{-\lambda_n^2(1+a)\overline{T}}{a}\right) \tag{5.26}$$

where the λ_n's are the roots of

$$\cot \lambda_n = \frac{\lambda_n}{a}$$

Permeable Top, Permeable Bottom (*PTPB*) (i).

$$m_2 = 0$$

Hence $a = \alpha = \infty$ but B is finite

i.e. layer 2 has a finite permeability and thickness but is incompressible. It therefore impedes the consolidation of layer 1 but does not contribute to the total settlement. In layer 1

$$\frac{\partial u}{\partial \overline{T}} = \frac{1+B}{B} \frac{\partial^2 u}{\partial y^2}$$

where depth from top is yh_1. With boundary conditions

$$(u)_{y=0} = 0, \qquad \left(\frac{\partial u}{\partial y}\right)_{y=1} = -B(u)_{y=1} \qquad \text{and} \quad (u)_{\overline{T}=0} = 1$$

solution is

$$u = 2 \sum_{n=1}^{\infty} \frac{(1-\cos \lambda_n)\sin(\lambda_n y)}{(\lambda_n - \sin \lambda_n \cos \lambda_n)} \exp\left(\frac{-\lambda_n^2(1+B)\overline{T}}{B}\right) \tag{5.27}$$

and $$U_s = 1 - 2 \sum_{n=1}^{\infty} \frac{(1-\cos \lambda_n)^2}{\lambda_n(\lambda_n - \sin \lambda_n \cos \lambda_n)} \exp\left(\frac{-\lambda_n^2(1+B)\overline{T}}{B}\right) \tag{5.28}$$

where the λ_n's are roots of

$$\tan \lambda_n = \frac{-\lambda_n}{B}$$

Permeable Top, Permeable Bottom (*PTPB*) (ii).

$$k_2 = \infty$$

Hence $\alpha = 0$ and $B = \infty$ but a is finite

i.e. layer 2 is compressible but fully permeable so that it does not impede the drainage of layer 1 but makes an immediate contribution to the total settlement of the whole deposit. Thus consolidation of layer 1 is governed by the ordinary Terzaghi theory for two-way drainage. Degree of settlement of whole deposit,

$$U_s = \frac{1 + aU_T}{1 + a} \tag{5.29}$$

where U_T is degree given by Terzaghi theory, the time factor being $4c_1 t/h_1^2$. Then

$$\overline{T} = \frac{\bar{c}t}{H^2} = \left(\frac{a}{1+a}\right)\frac{c_1 t}{h_1^2} \tag{5.30}$$

In *Figures 5.1* and *5.2* some numerical results are shown for a single pair of layers (marked ① in the figure) and for an infinite number of layer pairs (marked ∞). The latter curve is identical with that for a single homogeneous layer irrespective of the values of 'a' and 'α' and, furthermore, this particular curve is the curve for a single pair of layers when $\alpha = 1$. The curves in *Figure 5.1* (*PTPB*) also apply to values of 'a' and 'α' which are both the inverse of those specified in the figure, whereas inversion of these quantities for *Figure 5.2* (*PTIB*) renders the curves applicable to the drainage conditions (*ITPB*).

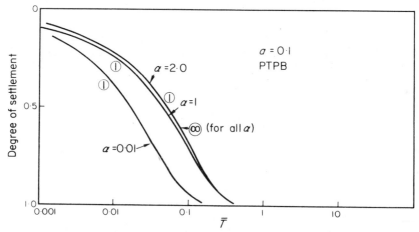

Figure 5.1   Rate of settlement for multiple layer pairs (1 one pair, ∞ infinite number of pairs). Permeable top, permeable base. (PTPB)

The presentation of the complete range of settlement-time curves is, for most situations, unnecessary. It is evident that, at least for degrees of settlement in excess of 0.2 for *PTPB* and 0.4 for *PTIB*, the curves for a single layer pair are geometrically similar in shape and are similar to that for the

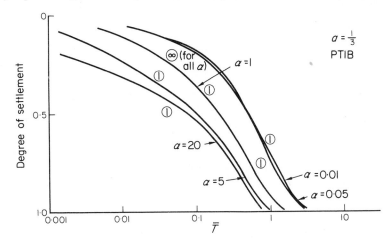

Figure 5.2   Rate of settlement for multiple layer pairs. Permeable top. Impermeable base. (PTIB)

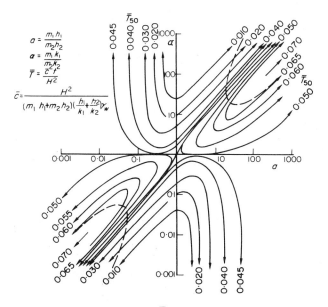

Figure 5.3   Contours of $\bar{T}_{50}$ for two layer soil (PTPB)

homogeneous layer. Thus it is sufficient to record the time for, say, 50% settlement and rely on the shape similarity to obtain the rest of the settlement-time curve. Contours of $\bar{T}_{50}$ are shown in *Figure 5.3* for *PTPB* and in *Figure 5.4* for *PTIB*. The plotted results cover the appropriate range of '*a*' and '*α*' (or '*B*').

For the condition *PTPB*, inversion of both 'a' and 'α' must lead to the same value of $\bar{T}_{50}$. The symmetry following from this requirement is evident in *Figure 5.3*. There is no corresponding symmetry in *Figure 5.4* (*PTIB*) but inversion of both 'a' and 'α' on this figure gives values of $\bar{T}_{50}$ for condition *ITPB*.

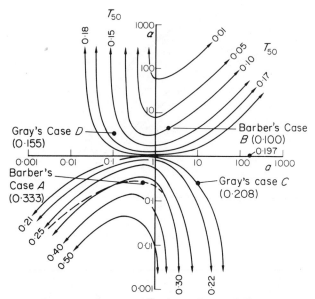

Figure 5.4   Contours of $\bar{T}_{50}$ for two layer soil (*PTIB*)

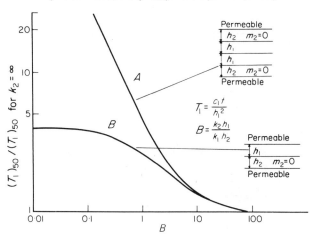

Figure 5.5   Oedometer test. Effect of imperfect porous plates

As recognised by Gray (1945), the limiting case *PTIB* (ii) is relevant to an oedometer test in which the soil specimen is loaded between two equally imperfect porous plates. Curve A of *Figure 5.5* gives the results of calculations for this limiting case in a form suitable for assessment of the effect of inadequate permeability of the porous plates. For example, if the thickness

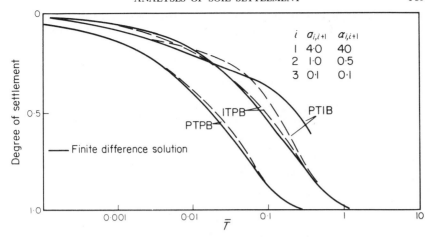

i	$\alpha_{i,i+1}$	$a_{i,i+1}$
1	4·0	40
2	1·0	0·5
3	0·1	0·1

Figure 5.6 Rate of settlement. Four layer deposit

of each plate is equal to half that of the specimen, the coefficient of permeability of the plates has to be less than twenty times that of the soil for the time for 50% consolidation to be increased by more than 20%.

Curve B in *Figure 5.5* has been calculated from limiting case $PTPB$ (i) and, in the present context, is relevant to an oedometer test with only one imperfect plate.

For deposits consisting of more than two disparate layers it is necessary to programme the finite difference or finite element (see Section 5.6) solutions. An example of the analysis for a four-layered deposit is shown in *Figure 5.6*.

5.3.2   Axially symmetric consolidation. Sand drains

Rowe and Horne showed that the presence of thin permeable layers would have a profound influence on the rate of consolidation of a soil deposit in which sand drains were installed. Some important aspects of the analysis were discussed in the earlier volume (Lee (1968)) and only some important subsequent results are discussed in this section.

Figure 5.7 shows the degree of settlement-time factor plots for a range of values of a parameter which expresses the geometry of the soil model. In this plot the time factor is expressed, for convenience, as

$$T_h = c_{h_1} \frac{(1+b/a)}{(1+1/a)} \cdot \frac{t}{4R_e^2} \tag{5.31}$$

the symbols were previously defined, but are again defined on the figure.

In the portion of the settlement-time factor curve geometrically similar to the settlement-time factor curve for a homogeneous layer, the equivalent coefficient of consolidation in the horizontal direction, $\bar{c}_r$, is independent of the degree of settlement U_s. Referring to *Figure 5.7* the ratio, y/z is equal to $\bar{c}_r/\bar{c}_h$, where,

$$\bar{c}_h = c_{h_1} \frac{(1+b/a)}{(1+1/a)} \tag{5.32}$$

The curves plotted in the figures are quite closely geometrically similar for U_s greater than 10–20% and this is more clearly shown by the relationships between $\bar{c}_h/\bar{c}_r$ and U_s, Figure 5.8.

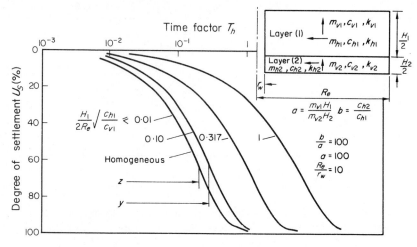

Figure 5.7   *Rate of settlement. Sand drain in layered soil*

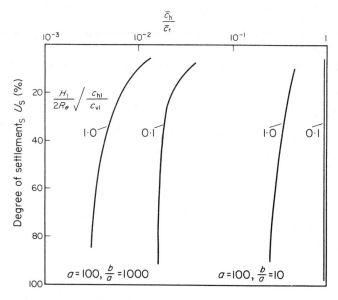

Figure 5.8   *Variation of equivalent horizontal coefficient of consolidation during consolidation process*

The effectiveness of the analysis can be readily established provided the parameters of the model can be defined (for example data by Rowe and Shields (1965)). This is, however, an extremely difficult task, and at the present time one of the major contributions of the analysis is the explanation of the need for testing large diameter specimens.

There are certain other theoretical implications worth considering. For example, when the layers are of comparable compressibility and thickness $\bar{c}_r$ will not be a constant during the consolidation process. *Figures 5.9* and *5.10* show the relationship between U_s and T_h for the relative compressibility parameter 'a' equal to unity. Large deviations from the shape for the

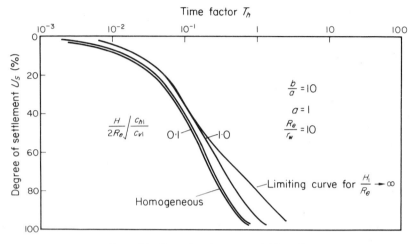

Figure 5.9   Rate of settlement. $a = 1$ $b/a = 10$

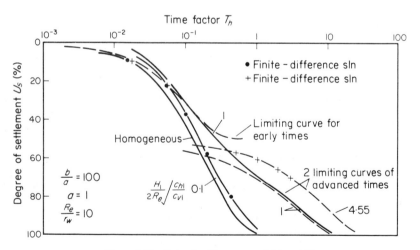

Figure 5.10   Rate of settlement. $a = 1$ $b/a = 100$

homogeneous soil increase as the ratio b/a increases. One interpretation is that the layer 2 is much more permeable than 1 so that, in the early stages, settlement is primarily due to consolidation of layer 2 in the direction of the layering, whereas in the later stages settlement is due to consolidation of layer 1 with pore pressures in layer 2 remaining effectively zero.

Thus for early times,

$$U_s \simeq \frac{U_{h_2}}{1+a} \tag{5.33}$$

where U_{h_2} is the degree of settlement due to flow in the horizontal direction of layer 2 alone. The associated time factor is

$$T_h = \frac{c_{h_2}t}{4R_e^2}\frac{(1/b+1/a)}{(1+1/a)} \tag{5.34}$$

For advanced times,

$$U_s = \frac{(1-(1-U_{h_1}).(1-U_{v_1}))a+1}{1+a} \tag{5.35}$$

where U_{h_1} and U_{v_1} are the degrees of settlement established by treating the consolidation flow in the direction of, and at right angles to, layering as independent processes.

The time factor is

$$T_h = T_{h_1}\frac{(1+b/a)}{(1+1/a)} \tag{5.36}$$

In these expressions T_{h_1} is the time factor associated with U_{h_1} and U_{v_1} is the degree of settlement for a time factor of

$$T_{v_1} = T_{h_1}\left(\frac{R_e^2 c_{v_1}}{c_{h_1}H_1^2}\right) \tag{5.37}$$

It is clear from *Figure 5.10* that the settlement-time relationships approach the special cases derived above. For $a = 1$ the curves for early time asymptote to a U_s value of 0.50. As a is increased the relative compressibility effect of the second layer diminishes and in *Figure 5.7*, for example, the relationship appears as a smooth curve asymptoting to zero U_s at small times. There are also other special cases which can be considered but are outside the scope of this section.

The solution to this problem can also be effected by finite difference (Lee (1968)) or finite element techniques (see Section 5.6). In *Figure 5.10* some numerical results of the finite difference analysis are plotted for comparative purposes. It should be noted however that the finite difference solution is inefficient with respect to computer time, and convergence problems are common. The evaluation of the formal solution, although tedious, is preferable and the finite element evaluation is satisfactory.

5.4   THREE-DIMENSIONAL CONSOLIDATION

Biot's theory (Biot (1941a), (1941b), (1955), (1963), Biot and Clingan (1941), (1942), Biot and Willis (1957)) was derived in the earlier volume (Lee (1968) p. 168). The primary purpose of the present section is to summarise the solutions now available for circular, square, strip and rectangular footings. It should be recalled that the Biot equation differs from the traditional diffusion equation by the term $\partial\Theta/\partial t$ where Θ is the mean total normal stress (p. 170), i.e.

$$\Theta = \sigma_{xx} + \sigma_{yy} + \sigma_{zz} \qquad (5.38)$$

and thus for a Poisson's ratio of $\frac{1}{2}$ the equations become identical since Θ is a constant. Thus the solution to the diffusion equation, using the appropriate c value† establishes a limiting relationship between the degree of settlement U_s and the time factor. A study of the numerical solutions shows that the value of Poisson's ratio does have a significant effect on the settlement-time relationship. Some early attempts were made to obtain simplified solutions, but the availability of exact evaluations has now obviated the need for the approximate approach.

The major variables involved in the solution for a surface foundation are

(1) Shape—circular, rectangular, square, strip.
(2) Permeability of footing, top and bottom of layer.
(3) Poisson's ratio of soil with respect to effective stresses.
(4) Thickness of layer.
(5) Roughness of base and footing.
(6) Nature of loading.
(7) Relative rigidity of footing.
(8) Anisotropy and non-homogeneity of layer.

Spectrums of solutions are now available covering the first four variables in some detail but the effects of anisotropy, layering, foundation rigidity, and type of loading has received little attention. The results quoted below were obtained by the evaluation of the Biot equations but the finite element analysis (Chapter 2, and Section 5.6) does offer an alternative solution which has the particular advantage of flexibility.

1. *Flexible Circular Footings Uniformly Loaded*
   (a) Semi-infinite, isotropic, homogeneous layer De Jong (1957).
   (b) Finite, isotropic, homogeneous layer, Gibson *et al.* (1970), Davis and Poulos (1971).

2. *Flexible Strip. Uniformly Loaded*
   (a) Semi-infinite, isotropic, homogeneous layer, Biot (1941b), Biot and Clingan (1942), McNamee and Gibson (1960), Schiffman *et al.* (1969).
   (b) Finite, isotropic, homogeneous layer, Gibson *et al.* (1970), Davis and Poulos (1971).

3. *Flexible Rectangle and Square. Uniformly Loaded*
   (a) Semi-infinite, isotropic, homogeneous layer, Gibson and McNamee (1957).

Mention should be made of the contribution of Mandel to the case of a clay layer overlain by a sand layer (Mandel (1957), (1961)) and the particularly significant analysis carried out by Cryer (1963) who established that the

† The usual subscript v is not used herein. c_1, c_2, c_3, are defined in terms of the effective elastic parameters; see Lee (1968), p. 153.

excess pore pressures could initially increase at some points within the consolidating layer.

Table 5.1 is a summary of the values of degree of consolidation of the centre of a uniformly loaded, flexible circular area. It will be noted that the solutions for a finite layer are rather limited except for the simplified situation of $v' = 0.5$.

Table 5.1 VALUES OF DEGREE OF CONSOLIDATION SETTLEMENT AT CENTRE OF A CIRCULAR, FLEXIBLE, FOUNDATION. UNIFORMLY LOADED.

Boundary conditions	v'	h/a	$c_1 t/a^2$ Semi-infinite layer $c_1 t/h^2$ Finite layer								Remarks
			10^{-4}	10^{-3}	10^{-2}	10^{-1}	0.50	1	5	10	
PT, PF	0	∞	0.02	0.04	0.11	0.35	0.63	0.73	0.87	0.91	Semi-infinite
	0.25	∞	—	—	0.14	0.42	0.63	0.77	0.89	0.92	layer
	0.50	∞	0.03	0.07	0.21	0.52	0.74	0.82	0.92	0.94	
PT, IF	0	∞	0	0	0.01	0.10	0.35	0.49	0.75	0.83	Semi-infinite
	0.50	∞	0	0.01	0.02	0.19	0.51	0.64	0.84	0.89	layer
PT, PB, PF	0.50	0.5	0.02	0.07	0.23	0.70	0.99	—	—	—	Finite layer
	0.50	2	—	0.15	0.37	0.85	0.99	—	—	—	Rough base
	0.50	10	—	0.44	0.75	0.94	0.99	—	—	—	
PT, IB, PF	0.50	0.5	—	0.03	0.11	0.38	0.78	0.95	—	—	Finite layer.
	0.50	2	—	0.06	0.25	0.95	—	—	—	—	Rough base
	0.50	10	0.09	0.33	0.66	0.96	—	—	—	—	
	0	1	0.01	0.03	0.12	0.40	0.80	0.92	—	—	Finite layer.
	0	10	0.11	0.37	0.76	0.96	—	—	—	—	Smooth base
PT, IB, IF	0.5	0.5	—	0.03	0.11	0.37	0.92	—	—	—	Finite layer.
	0.5	2	—	0.03	0.15	0.64	0.99	—	—	—	Rough base
	0.5	10	0.01	0.12	0.64	0.96	—	—	—	—	
PT, IB, IF	0.5	0.5	—	—	—	0.01	0.19	0.38	0.92	—	Finite layer.
	0.5	2	—	—	0.05	0.45	0.91	0.99	—	—	Rough base
	0.5	10	0.01	0.12	0.63	0.93	—	—	—	—	

In this table a is the radius, h is the thickness of the layer and c_1, is defined as

$$c_1 = \frac{k}{\gamma_w} \frac{E(1-v')}{(1+v')(1-2v')}$$

that is, the coefficient of consolidation corresponding to a one-dimensional state.

Diffusion equation solutions ($v' = \frac{1}{2}$) obtained by Davis and Poulos (1971). Other solutions due to other authors quoted in this section.

For practical purposes the solution for a square foundation can be obtained from the rate of settlement of a circular foundation of equal area.

When a strip is supported on a semi-infinite mass the final consolidation settlement is theoretically infinite, and it is necessary to define degree of settlement as the degree of differential settlement. Alternatively, the consolidation settlement at any time can be directly evaluated from the Biot equations (McNamee and Gibson (1960)).

Recently solutions for a strip on a finite layer have been evaluated and *Table 5.2* details the degree of consolidation settlement at the centre of a flexible strip subjected to a uniform pressure.

In *Table 5.2* a is the half width of the strip. Values quoted for $v' = 0.5$ were published by Davis and Poulos (1971), and values quoted for $v' = 0$ are due to Gibson and co-workers (McNamee and Gibson (1960), Gibson et al. (1970)). An examination of the published data shows that the effect of the layer depth is significant for h/a values less than 10.

Table 5.2 VALUES OF DEGREE OF CONSOLIDATION SETTLEMENT AT CENTRE OF A FLEXIBLE STRIP UNIFORMLY LOADED

Boundary conditions	v'	h/a	10^{-4}	10^{-3}	10^{-2}	10^{-1}	0.5	1	10	Remarks
PT, IB, PF	0	0.2	—	—	0.05	0.20	0.55	0.77	0.99	Finite layer
	0	1	—	0.02	0.08	0.29	0.70	0.87	0.99	
	0	5	0.02	0.09	0.29	0.60	0.92	0.99	—	
	0	10	0.05	0.17	0.42	0.74	0.96	0.99	—	
PT, IB, PF	0.5	1	—	0.04	0.13	0.42	0.84	0.95	—	Finite layer.
	0.5	5	—	0.09	0.26	0.63	0.92	0.98	—	Rough base
	0.5	10	—	0.14	0.31	0.67	0.94	0.99	—	
PT, PB, PF	0.5	1	—	0.07	0.23	0.73	0.98	—	—	Finite layer.
	0.5	5	0.03	0.16	0.40	0.83	0.99	—	—	Rough base
	0.5	10	0.05	0.21	0.47	0.87	0.99	—	—	
PT, PB, IF	0.5	1	—	—	0.12	0.41	0.85	0.96	—	Finite layer.
	0.5	5	—	0.05	0.18	0.70	0.98	—	—	Rough base
	0.5	10	0.02	0.07	0.27	0.81	0.99	—	—	
PT, IB, IF	0.5	1	—	—	—	0.06	0.36	0.55	—	Finite layer.
	0.5	5	—	—	0.09	0.51	0.84	0.93	—	Rough base
	0.5	10	—	0.02	0.18	0.65	0.9	—	—	

(header: $\dfrac{c_1 t}{h^2}$)

Figure 5.11 shows the degree of consolidation of a strip on a finite layer as a function of the time factor. These are results derived by Gibson (Gibson et al. (1970)) and quoted by Hwang et al. (1971) in their discussion of the effectiveness of the finite element solution of the Biot equation, the ratio of half width to depth of layer being unity, $v' = 0$ and the relationship applied to the centre line. G is the shear modulus, thus the coefficient of consolidation in this figure is related to c_1 by the equation

$$c_1 = \bar{c}\left(\frac{1-v'}{1-2v'}\right)$$

It will be noted that the time factor is defined as $\bar{c}t/a^2$ in order to directly illustrate the effect of layer thickness on rate of consolidation.

It is possible to interpolate values of U_s for v' intermediate between the quoted limits of 0 and 0.5. The effect of an increase in v' is to increase the rate of settlement.

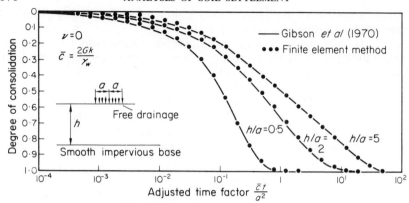

Figure 5.11   Rate of consolidation for a strip load (after Hwang et al. (1971). Courtesy Canadian Geot. Jour.)

Gibson and McNamee's results for a degree of settlement at a corner of a rectangular footing were earlier plotted with the time factor equal to $c_3 t/BL$ (Lee (1968), p. 172, Davis and Poulos (1971)) and it was seen that the plot was not sensitive to the aspect ratio L/B. By superposition the degree of settlement at other parts can be readily determined.

5.5   ANALYSES OF RAFT FOUNDATIONS

There are three basic elements which should be considered in the analysis of a raft foundation

(1) The structure.
(2) The foundation.
(3) The supporting soil.

It is common to neglect the effect of the structure in differential settlement calculations but it is evident that the stiffness of the structure can have a profound effect on the distribution of settlements and column loads and the magnitude of the maximum moment induced in the raft. The complete structure-foundation-supporting soil analysis is developed in this section, but before considering this interaction analysis in more detail it is necessary to pay some attention to existing methods of analysis and some recently developed analyses of the foundation-supporting soil system.

A satisfactory analysis of the latter system should take into account the three-dimensional nature of the problem. It may be expedient to reduce the problem to a one-dimensional state by adopting a simple model of the soil combined with the elementary beam theory. The soil-line method (Baker (1957)) is an analysis of this type in which the soil is represented by the Winkler model. If the foundation is rigid relative to the stiffness of the supporting soil, the contact pressure distribution becomes determinate whether the foundation is treated as a (one-dimensional) beam or a (two-dimensional) plate. This is commonly used in design and is referred to as the conventional method.

The more general analysis of a flexible raft has been considered by several investigators (for example, Murphy (1935), Gorbunov-Possadov (1959), Severn (1966), Zienkiewicz and Cheung (1967)) and the special cases of an infinite (Winkler and linear elastic) or semi-infinite (Winkler) have been solved (Holl (1938), Hogg (1938), (1943), Westergaard (1942)). The finite element method has the particular advantages that orthotropy and local stiffness variations in the raft can be readily taken into account. It is also possible, for example, to analyse a raft-pile system with only minor modifications to the programme.

It is reasonable to anticipate that the linear elastic model is a superior physical representation to the Winkler model. The only advantage of the use of the Winkler model is one of computational simplicity. For very flexible rafts the differences in predicted performance by the use of the alternative models is insignificant.

5.5.1 Foundation-supporting soil analysis

In this section three methods for dealing with raft foundations are reviewed. The first two are different approaches using the Winkler model and the third is the linear elastic continuum approach.

Winkler foundations: Method 1. The Winkler foundation model is characterised by the equation

$$p = k . w \qquad (5.39)$$

where p = pressure;

   w = deflection;

   k = modulus of subgrade reaction.

Figure 5.12 shows portion of the finite element grid representation of a rectangular plate resting on a Winkler-type support. If we consider any

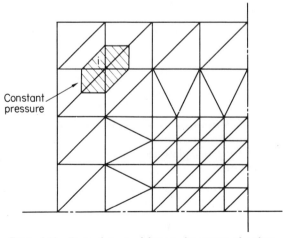

Constant pressure

Figure 5.12   Finite element subdivision for quarter of a plate

node, say i, then the deflection that results from a load applied at this node is equal to the deflection obtained when this load is distributed over the shaded portion. The area around node i is defined by joining the mid-points of all element boundaries that pass through the node. For the particular case of a rectangular grid $a \times b$ over the whole area then

$$\frac{R_i^*}{ab} = k \cdot w_i \qquad (5.40)$$

where R_i^* = foundation reaction force

$$\therefore \ R_i^* = abk \cdot w_i \qquad (5.41)$$

and hence for every node in the plate we have

$$\{R^*\} = abk[\beta] \cdot \{w\} \qquad (5.42)$$

where $[\beta]$ is a diagonal matrix with terms

$$1 = i \text{ is interior pt.}$$
$$\tfrac{1}{2} = i \text{ is edge pt.}$$
$$\tfrac{1}{4} = i \text{ is corner pt.}$$

This representation of the foundation does not contain any terms for the rotation stiffness of a node. Before it can be assembled into the plate stiffness matrix it is necessary to augment so that there are as many terms as in the plate matrix. Hence,

$$\{Q^*\} = [K_F] \cdot \{\delta\} \qquad (5.43)$$

Q^* = foundation reaction moments and forces for each node (moments by definition are zero).

$[K_F]$ = foundation stiffness matrix (actually the augmented $abk \ [\beta]$ matrix).

$\{\delta\}$ = nodal displacements.

Now, $\{R\} = \{Q\} - \{Q^*\}$

$\{Q\}$ = externally applied nodal forces.

$\{R\}$ = resultant forces which cause bending of the plate.

$$\therefore \ \{Q\} = \{R\} + \{Q^*\}$$
$$= [K_p + K_F]\{\delta\} \qquad (5.44)$$

where, $[K_p]$ = the plate stiffness matrix.

Winkler foundations: Method 2.   Whereas the first method was based purely on statical requirements this approach is based on strain energy principles. In formulating the plate element stiffness matrix the derivation was based on the equation

$$\frac{\text{strain energy of}}{\text{nodal point system}} = \frac{\text{strain energy of}}{\text{the stress-strain system}}$$

If however the plate is supported by a Winkler foundation then it is necessary to consider the contribution to the strain energy of the system that comes from deformation of the foundation. Hence,

$$\frac{\text{strain energy}}{\text{nodal point system}} = \frac{\text{strain energy}}{\text{stress-strain system}} + \frac{\text{strain energy}}{\text{foundation deformation}}$$

$$\frac{\text{strain energy}}{\text{foundation deformation}} = \tfrac{1}{2} \int_A p.w.dA$$

$$= \tfrac{1}{2} \int_A kw^2.dA \qquad (5.45)$$

$w = $ the vertical deflection.

Polynomial representation for w can be chosen as

$$w = [M(x, y)] \{\alpha\} \qquad (5.46)$$

$M(x, y) = $ terms in the polynomial;

$\{\alpha\} = $ unknown coefficients of the terms.

Now, w^2 is defined as $w^T.w$ so the foundation term becomes

$$= \tfrac{1}{2} \int_A k.\alpha^T.M^T.M.\alpha.dA \qquad (5.47)$$

$$\{\alpha\} = [C^{-1}].\{v\} \qquad (5.48)$$

where $v = $ nodal displacements for the element;

$C = $ transformation matrix.

$$\frac{\text{Strain energy}}{\text{foundation deformation}} = \tfrac{1}{2}[C^{-1}]^T.k.\int_A M^T.M \, dA.[C^{-1}] \qquad (5.49)$$

This gives the expression for the stiffness matrix for a plate element on a Winkler foundation.†

$$|k^e| = [C^{-1}]^T \left[\int_A B^T.D.B.dA + k.\int M^T.M.dA \right] [C^{-1}] \qquad (5.50)$$

This expression is for continuous spring support over the whole element not just at the nodes. As a result, this approach does give better convergence than the previous method. This is particularly evident when we are considering plates of high flexibility. The results (*Figure 5.13*), of two flexibilities are shown firstly, the rigid plate where we can get an exact solution from statics, and secondly, a very flexible plate which we can compare with the deflection of an infinite plate of the same flexibility

Semi-infinite elastic continuum.   The deflection of any point j due to a point load at i on an isotropic elastic half-space is given by the Boussinesq equation

$$w_{ji} = \frac{1-v_0^2}{\pi E_o}.\frac{P_i^*}{r_{ij}} \qquad (5.51)$$

where, $P_i^* = $ load at i;

$r_{ij} = $ radial distance from i to j;

$E_o = $ Young's modulus of the foundation soil;

$v_o = $ Poisson's ratio of the foundation soil.

† Zienkiewicz' (1971) notation for B, D, and k^e.

180

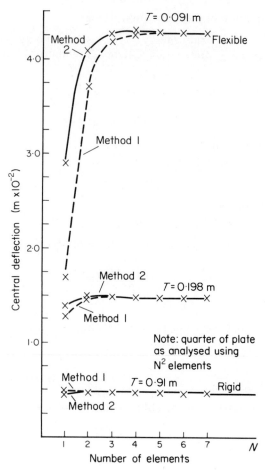

Figure 5.13   Convergence of plate solution. Winkler soil model central concentrated load (after Hain and Lee (1973))

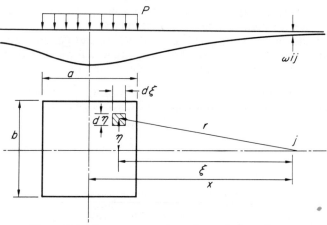

Figure 5.14   Element supported on a linear elastic medium

In this treatment it is convenient to assume a constant pressure acting on an area surrounding each nodal point. If we have a regular rectangular grid of 'a' units by 'b' units then if i is an interior node this pressure will have a magnitude of P_i^*/ab and will vary from node to node. The deflection at node j caused by the uniformly loaded rectangular area $a \times b$ around node i can be obtained by integrating the Boussinesq equation over the rectangular area (Figure 5.14).

$$w_{ij} = \int_{\xi=-a/2}^{\xi=a/2} \int_{\eta=-b/2}^{\eta=b/2} \frac{P_i(1-v_0^2)}{ab\pi E_o} \cdot \frac{d\xi \cdot d\eta}{\sqrt{(\xi^2+\eta^2)}} \tag{5.52}$$

The limits of the integration will change if i is a corner node or an edge node and so for any set of rectangular grid points the deflections can be written as

$$\{w\} = \frac{1-v_0^2}{\pi E_o} \cdot [f_f] \cdot \{P^*\} \tag{5.53}$$

where $[f_f]$ = foundation flexibility matrix;

   $\{P^*\}$ = foundation reaction forces.

Inverting, one can write

$$\{P^*\} = \frac{\pi E_0}{(1-v_0^2)} \cdot [K_f] \cdot \{w\} \tag{5.54}$$

where $[K_f] = [f_f]^{-1}$.

This matrix has now to be combined with that of a plate subdivided into finite elements. Such a matrix connecting nodal forces Q and displacements δ is

$$\{Q\} = [K_p] \cdot \{\delta\} \tag{5.55}$$

For each nodal force Q_i and displacement δ_i three components are present. These correspond to vertical displacement w_i and two rotations θ_{xi} and θ_{yi}. As no angular continuity is assumed between the foundation and the plate the rotations and corresponding moments can be eliminated by partial inversion from the above relation. If we rewrite as

$$\begin{pmatrix} F \\ M \end{pmatrix} = \begin{pmatrix} K_{11} & K_{12} \\ K_{21} & K_{22} \end{pmatrix} \cdot \begin{pmatrix} w \\ \theta \end{pmatrix}$$

then, $\{F - K_{12} \cdot K_{22}^{-1} \cdot M\} = [K_{11} - K_{12} \cdot K_{22}^{-1} \cdot K_{21}] \cdot \{w\} \tag{5.56a}$

or more succinctly,

$$\{P\} = [K_p^1] \cdot \{w\} \tag{5.56b}$$

where, $P = F - K_{12} \cdot K_{22}^{-1} \cdot M$

$K_p^1 = K_{11} - K_{12} \cdot K_{22}^{-1} \cdot K_{21}$

Noting that if W_i represents an externally applied vertical load to a node then $W_i - P_i^*$ is the effective external force acting on the node and we can write for the plate,

$$\{W\} - \{P^*\} = [K_p^1] \cdot \{w\} \tag{5.57}$$

Hence,
$$\{W\} = \left[K_p^1 + \frac{\pi E_0}{1-v_0^2} \cdot K_f \right] \cdot \{w\} \qquad (5.58)$$

A direct solution of this equation will give deflections which in turn will give rotations and contact pressures and the bending moments in the plate.

It will be noticed that this approach is considerably different from the one that was used in the Winkler method 1 solution. The reason for this will become apparent if we consider the foundation stiffness matrix for each case. For the Winkler case this matrix is a diagonal one of order (number of nodes) but for the half-space it is a fully populated matrix of order n. If we augment each of these matrices with zeros for the rotational terms we have,

$$\text{storage for Winkler} \ = 3n \times 1$$

$$\text{storage for half-space} = (3n+1) \times 3n/2$$

where n = number of nodes.

Hence the large amount of storage needed for the half-space stiffness matrix would severely limit the size of problem which can be solved.

Another difference between the Winkler solution and the half-space solution that is worth mentioning is the modification necessary to take into account the symmetry of a problem. As far as the plate is concerned we must specify

$$\frac{\partial w}{\partial (\text{normal})} = 0 \qquad (5.59)$$

along all the lines of symmetry. The Winkler foundation requires no modification but with the half-space model account must be taken of the effect of

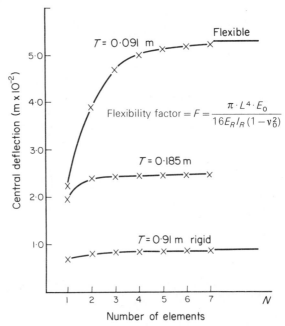

Figure 5.15   Convergence of plate solution elastic half space central concentrated load

loads which act in the other portion of the foundation.

$$\begin{pmatrix} w_1 \\ w_2 \end{pmatrix} = \frac{1-v_0^2}{\pi E_0} \begin{bmatrix} f_{11} & f_{12} \\ f_{21} & f_{22} \end{bmatrix} \cdot \begin{pmatrix} P_1^* \\ P_2^* \end{pmatrix} \tag{5.60}$$

which is a partitioned form of equation 5.53.

For single line of symmetry $P_1^* = P_2^*$ hence

$$\{w_1\} = \frac{1-v_0^2}{\pi E_0} [f_{11}+f_{12}] \cdot \{P_1^*\} \tag{5.61}$$

hence,
$$[K_f] = [f_{11}+f_{12}]^{-1} \tag{5.62}$$

Results of convergence tests for several plate thicknesses are shown in *Figure 5.15*. We notice that the flexible plate $T = 0.091$ m is considerably slower to converge than the thick plate $T = 0.91$ m. This is because with the flexible plate, the equal pressure area around each node is not as good an approximation as with the rigid plate.

Let us now consider a practical situation in which a raft appears to be relatively flexible (0.61 m deep), and assume that the differential settlements do not cause a major redistribution of the loads applied to the raft, that is, the structure is flexible. Portion of the raft may be considerably stiffened by lift wells, and in the layout of elements for both the Winkler and linear elastic models the stiffness of the elements in a lift-well area is increased. If the real situation of infinite stiffness was modelled some difficulties arise because the stiffness matrix becomes ill-conditioned.

At this stage of the analysis we will not take into account the actual dimensions of the columns, but simply assume that the loads are applied at the nodal points.

One of the important results of the analysis is the distribution of contact pressure along, say, a line of columns. *Figure 5.16* shows such a distribution for the Winkler model, and the corresponding results for the linear elastic model are shown in *Figure 5.17*. We can see by a comparison with the distribution for an infinite plate, that the latter happens to be very close to the finite element solution within the interior of the plate (but not, of course, in the vicinity of the lift wells).

Although the pressures and displacements are accurately determined by the analyses this is not true with respect to the bending moments in the raft.

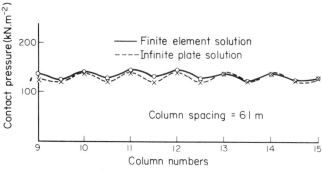

Figure 5.16   Contact pressure distribution along interior line of columns. Raft 0.61 m thick. Plan dimensions 36.6 m × 18.3 m. Column spacing 6.1 m. Winkler soil model

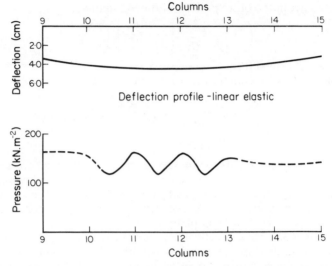

Figure 5.17   *Deflection and contact pressure distributions along interior line of columns. Linear elastic soil model*

To establish the moments to within an acceptable accuracy it is usually necessary to adopt a higher order element, or if there is sufficient computer capacity one may simply increase the number of the rectangular elements. It happens, however, that such procedures are not necessary in the present situation because of the high relative flexibility of the raft. One can consider a typical interior region and approximate quite closely to the real situation

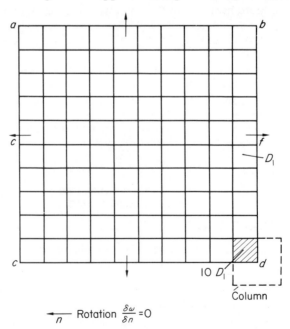

Figure 5.18   *Local subdivision of elements*

by adopting the boundary conditions depicted in *Figure 5.18*. Such conditions are justified by the type of interaction between the columns.

The elements can be concentrated within a square of dimensions equal to half the column spacing thus defining the local moments with adequate accuracy. Furthermore, we can treat the column as a square of finite dimensions with a stiffness well in excess of the raft stiffness.

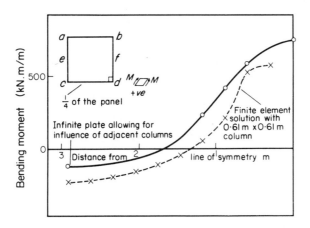

Figure 5.19   Bending moment profile along cd.

Figure 5.19 shows the distribution of bending moment along the line of columns considering that the column is 0.61 m square. Again, a comparison can be made with an infinite plate. However, it is necessary in the latter case to assume that the column load is uniformly distributed over the area of the column.

5.5.2   Interaction analysis

Following the development of the analysis of the foundation-supporting soil system, we can now proceed to a consideration of the effect of the stiffness of the superstructure on the behaviour of the raft. There are several possible techniques for 'attaching' the structure to the raft. In the earlier studies (Lee and Harrison (1970), Lee and Brown (1972)) the stiffness characteristics of the structure were first established for zero settlements (pinned ends), then the sway moments, and differential settlement effects were introduced by satisfying displacement and rotational compatibility at the column-raft intersections. An alternative procedure, which has mathematical appeal from the point of view of procedural consistency, is the sub-structures approach (Hain and Lee (1973)). In this method the stiffness of the structure is directly incorporated into the stiffness matrix of the raft.

Let us now consider the sub-structure method in more detail.

The complete set of equilibrium equations for frame regarded as a free body may be written in matrix form as

$$\{P\} = [K].\{U\} \qquad (5.63)$$

Let, U_b = the displacements at each of the lower column bases, i.e. boundary displacements common to the frame and the raft;

U_i = all other displacements of the frame, i.e. interior displacements which only occur in the frame.

The corresponding effective external forces are denoted by P_b and P_i. Hence we can rewrite equation 5.63 in partitioned form as

$$\begin{bmatrix} K_{bb} & K_{bi} \\ K_{ib} & K_{ii} \end{bmatrix} \cdot \begin{pmatrix} U_b \\ U_i \end{pmatrix} = \begin{pmatrix} P_b \\ P_i \end{pmatrix} \tag{5.64}$$

It is now assumed that the total displacements of the structure can be calculated from the superposition of two matrices such that

$$\{U\} = \{U^\alpha\} + \{U^\beta\} \qquad \text{(Przemieniecki (1968))} \tag{5.65}$$

where $\{U^\alpha\}$ = displacements due to P_i with $U_b = 0$;

$\{U^\beta\}$ = corrections to $\{U^\alpha\}$ to allow for boundary displacements U_b with $P_i = 0$.

$$\{U\} = \begin{pmatrix} U_b \\ U_i \end{pmatrix} = \begin{pmatrix} U_b^\alpha \\ U_i^\alpha \end{pmatrix} + \begin{pmatrix} U_b^\beta \\ U_i^\beta \end{pmatrix} \tag{5.66}$$

where by definition,

$$\{U_b^\alpha\} = 0$$

Similarly the external forces can be separated into

$$\{P\} = \{P^\alpha\} + \{P^\beta\}$$

$$\begin{pmatrix} P_b \\ P_i \end{pmatrix} = \begin{pmatrix} P_b^\alpha \\ P_i^\alpha \end{pmatrix} + \begin{pmatrix} P_b^\beta \\ P_i^\beta \end{pmatrix} \tag{5.67}$$

where by definition,

$$\{P_i^\alpha\} = \{P_i\}$$
$$\{P_i^\beta\} = 0$$

From equation 5.64 if $U_b = 0$, we have

$$\{U_i^\alpha\} = [K_{ii}^{-1}] \cdot \{P_i\} \tag{5.68}$$

and,

$$\{P_b^\alpha\} = [K_{bi} \cdot K_{ii}^{-1}] \cdot \{P_i\} \tag{5.69}$$

$\{P_b^\alpha\}$ represents boundary reactions necessary to maintain $U_b = 0$ and hence it is the column loads and moments for fixed columns with zero differential settlement. These are the forces which would normally be applied to the foundation to calculate moments and shears.

Also,

$$\{U_i^\beta\} = -[K_{ii}^{-1} \cdot K_{ib}] \cdot \{U_b^\beta\} \tag{5.70}$$

$$[K_{bb} - K_{bi} \cdot K_{ii}^{-1} \cdot K_{ib}] \cdot \{U_b^\beta\} = -[K_{bi} \cdot K_{ii}^{-1}] \cdot \{P_i\} \tag{5.71}$$

hence, $[K_{bb} - K_{bi} \cdot K_{ii}^{-1} \cdot K_{ib}]$ is the contribution of the frame to the total stiffness of the whole system frame plus raft plus soil.

Now we have to consider the raft plus soil of the system. Using one of the methods outlined previously we can represent the complete set of equilibrium equations for the raft and soil as

$$\{F\} = [K_f] \cdot \{\delta\} \tag{5.72}$$

The boundary displacements stiffness matrix for the frame is now merged into K and the boundary reactions into F.

$$\{F\} = [K_T].\{\delta\} \tag{5.73}$$

This equation can be solved for $\{\delta\}$

$$\{\delta\} = [K_T^{-1}].\{F\} \tag{5.74}$$

The boundary displacements $\{U_b^\beta\}$ for the frame can now be extracted from $\{\delta\}$.

Now, $$[K_{bb} - K_{bi}.K_{ii}^{-1}.K_{ib}].\{U_b^\beta\} = \{\bar{S}_b\} \tag{5.75}$$

$\{\bar{S}_b\}$ are the column forces and moments when the structure and raft are analysed as a compatible unit.

On back substitution for the structure

$$U = \begin{pmatrix} U_i \\ U_b \end{pmatrix} = \begin{pmatrix} K_{ii}^{-1}.P_i - K_{ii}^{-1}.K_{ib}.U_b^\beta \\ U_b^\beta \end{pmatrix} \tag{5.76}$$

Using U the member end forces for any element in the frame can be calculated.

As an example of the interaction analysis let us consider a 3 bay by 3 bay frame; we will concentrate on the influence of the relative stiffness of the raft and the stiffness of the structure.

Figure 5.20 shows the variation in column load as a function of the relative flexibility parameter, λL, defined as,

$$L = 4\sqrt{\left(\frac{kL^4}{4E_R I_R}\right)} \tag{5.77}$$

where L = length of the raft;

$E_R I_R$ = flexural rigidity of the raft/unit width;

k = Winkler stiffness of the supporting soil.

Column 1 of *Figure 5.20* is a corner column, columns 2 and 3 are at the edge of the raft, and 4 is an interior column. The loads for a rigid raft are the values corresponding to $\lambda L = 0$. When a comparison is made with an identical structure supported on a raft of reduced stiffness (or on a soil of increased k), it is seen that there is a transfer of load from columns 1, 2, and 3 to the interior column. This transfer continues progressively with increasing raft flexibility. Limiting values, calculated by a completely independent analysis, are shown in *Figure 5.20* as short lines and it is evident that the loads established by the interaction analysis asymptote to these limiting values.

The terms 'pinned' and 'fixed' refer to the fixity condition between the columns and the raft, that is, moment-free or built-in to the raft. The effect of column fixity is shown to be quite small when the relative flexibility is low. There is approximately a 15% difference of column loads at values of λL in excess of, say, 3. Fixity tends to reduce the changes induced in the column loads as the raft flexibility is increased.

It is difficult to define the actual stiffness of a practical structure, and in *Figure 5.20* the effect of structural stiffness is illustrated. Here we are considering the column loads developed when the structure is so stiff that the

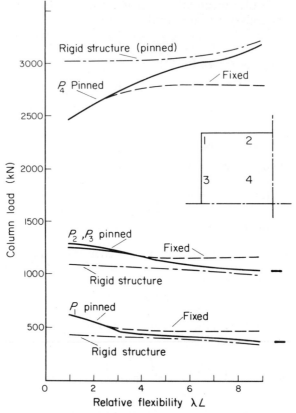

Figure 5.20   Interaction analyses. Variation of column loads with relative flexibility of foundation Winkler model. Three bay space frame (after Hain and Lee (1973))

differential settlements are negligible—referred to as a 'rigid' structure. A comparison of the column loads for this case and the values obtained by the interaction analysis considering the frame stiffness alone, shows the range of column loads is greatest at low raft flexibilities. These 'limits' merge as the flexibility of the raft is increased.

The moments induced in the raft are of primary importance to the foundation engineer and *Figure 5.21* shows the maximum positive and negative moments as a function of relative flexibility. Maximum bending moments corresponding to the case when the column loads are unaffected by the differential settlement are shown in *Figure 5.21* as dotted lines, and marked 'constant column loads'. These moments are the values which would be calculated by the extension of the soil line method to two dimensions.

Now let us consider the effect of adopting the linear elastic model of soil behaviour. *Figures 5.22* and *5.23* are the results corresponding to *Figures 5.20* and *5.21* respectively when the raft is supported on a semi-infinite linear elastic continuum. There is quite a contrast in the predicted behaviour as seen by a study of the corresponding figures. The basic differences arise from the fact that the raft on a linear elastic material manifests a concave settlement pattern whereas the settlement profile of a raft on a Winkler material is

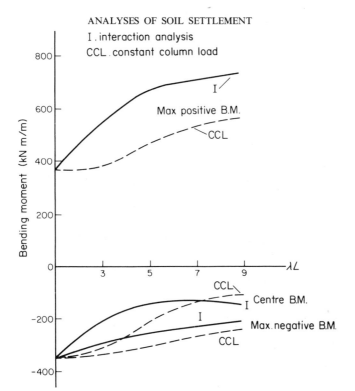

Figure 5.21   Effect of relative flexibility on bending moments. Winkler model. (after Hain and Lee (1973))

convex. Thus, the outer column loads increase in the case of a linear elastic supporting soil instead of decreasing as predicted by the analysis incorporating the Winkler model. Furthermore, the maximum positive moments decrease and increase, respectively, as the relative flexibility of the raft is increased.

It is considered that the linear elastic analysis should be used because of the superior physical (but far from complete) representation of a soil mass. The Winkler should only be applied to very flexible rafts. Used intelligently (compare with settlement) with predictions based on linear elastic model (Lee (1968), Chapter 3) the interaction analysis provides the geotechnical engineer a very useful analytical tool.

Advances in this analysis are being made which take into account nonlinear behaviour of soil mass and the non-homogeneous situation common in soil deposits.

5.6   FINITE ELEMENT ANALYSIS OF CONSOLIDATION PROBLEMS

The consolidation of soils as a result of the dissipation of excess pore pressures is considered important in order to achieve a better understanding of the behaviour of foundations and to predict more accurate settlements. The principal difficulties involved in any analysis of consolidation problems are

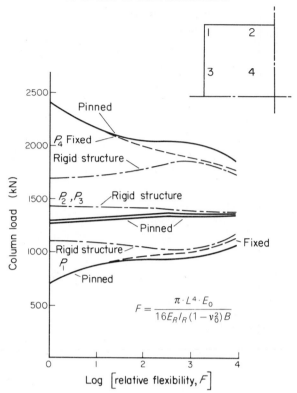

Figure 5.22   *Interaction analyses variation of column loads with relative flexibility of foundation. Linear elastic model. Three bay space frame. (after Hain and Lee (1973))*

the various mathematical complexities introduced in a more realistic formulation of the problem. The problem may be solved on the basis of one-, two- or three-dimensional theory of consolidation but in reality most of these problems fall into the category of three-dimensional consolidation. Again the relation between the effective stress and strain may be considered linear for the sake of simplicity whereas the investigation of non-linear behaviour during consolidation will represent a more common situation encountered in soil mechanics. Many investigations have been carried out to study the consolidation problem on the basis of one-, two- or three-dimensional nature (Terzaghi (1943), Biot (1941a), Schiffman *et al.* (1969)) and on the basis of non-linear behaviour (Davis and Raymond (1965), Gibson *et al.* (1967), Janby (1965)). However most of the solutions available to the consolidation problem are applicable only to the trivial situations due to the complexities such as non-homogeneous soil mass, non-linear material behaviour, anisotropic properties, three-dimensional nature of the problem and difficult representation of the actual boundary conditions.

Fortunately, many of the difficulties previously encountered in the analysis of the consolidation problem have now been reduced to a great extent due to the major expansion of the high-speed computers and the development of powerful numerical techniques such as the finite element method. Recently the finite element formulation based on the variational principle has been

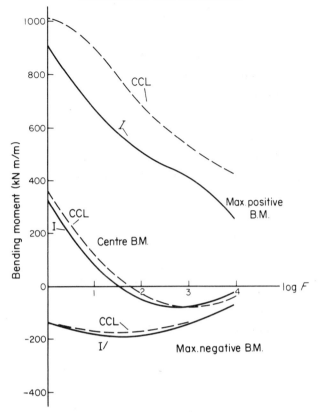

Figure 5.23   Effect of relative flexibility on bending moments linear elastic model (after Hain and Lee (1973))

proposed by Sandhu and Wilson (1969) and the procedure has been extended to include a logarithmic time increment by Hwang *et al.* (1971). An alternate procedure using the finite element method for plane strain consolidation has been developed by Christian and Boehmer (1970).

In this section the finite element analysis of three-dimensional consolidation based on the work of Sandhu and Wilson (1969) and Hwang *et al.* (1971), is described and the application of the method to the problem of strip load on elastic layer is discussed.

5.6.1   Field equations: Variational principle

The functional which is to be minimised for three-dimensional consolidation, in the case of initially undeformed soil, saturated with an incompressible pore fluid and subjected to small deformations with linear material behaviour, is defined as

$$\Omega_t(u, p) = \int_v \left[\tfrac{1}{2}\sigma_{ij}u_{i,j} - \rho F_i u_i + p u_{i,i} - \tfrac{1}{2}g^* q_i(p_i + \rho_2 F_i) \right] dv$$

$$- \int_{S1} (T_i u_i)ds + \int_{S2} (gQp)ds \qquad (5.78)$$

where u_i = displacements;

   p = pore water pressure,

   σ_{ij} = components of effective stress tensor;

   F_i = components of body force vector;

   ρ = mass density of soil;

   q_i = components of flux vector;

   ρ_2 = mass density of pore fluid;

   T_i = prescribed tractions on the surface $S1$;

   Q = prescribed flow normal to the surface $S2$;

   $g = 1$.

The boundary conditions to be satisfied are

$$\text{(a)} \quad T_i = (\sigma_{ij} + \delta_{ij}p)n_j \quad \text{on } S1 \qquad (5.79a)$$

where δ_{ij} = Kronecker delta;

   n_j = the direction cosines of the outward normal to the boundary.

and $$\text{(b)} \quad Q = q_i n_i \quad \text{on } S2 \qquad (5.79b)$$

5.6.2   Finite element idealisation

The domain is divided into a finite number of subregions called 'elements' which are interconnected by points known as 'nodes'. The displacement and the pore pressure at any point within an element are uniquely defined in terms of its nodal values.

If $\{u\}$ is the displacement vector and p is the pore pressure at any point, then

$$\{u\} = [N_u]^T\{\delta\}^e \qquad (5.80a)$$

$$p = [N_p]^T\{p\}^e \qquad (5.80b)$$

where N = the shape function defined by the interpolating polynomial;

   $\{\delta\}^e$ = the displacements u and v at the nodes of the element;

   $\{p\}^e$ = the pore pressure at the nodes of the element.

The interpolating polynomial may be linear, quadratic or cubic depending on the number of nodes on each side of the element.

The strain at any point can be written as

$$\{\varepsilon\} = \begin{bmatrix} \varepsilon_x \\ \varepsilon_y \\ \gamma_{xy} \end{bmatrix} = \begin{bmatrix} \dfrac{\partial u}{\partial x} \\ \dfrac{\partial v}{\partial y} \\ \dfrac{\partial u}{\partial y} + \dfrac{\partial v}{\partial x} \end{bmatrix} = [B_u]\{\delta\}^e \qquad (5.81)$$

where the transformation matrix $[B_u]$ is obtained from the displacement interpolation functions by appropriate differentiation.

Similarly the pore pressure gradient can be obtained by differentiating equation 5.80b

$$\{grad.\ p\} = \begin{bmatrix} \dfrac{\partial p}{\partial x} \\ \dfrac{\partial p}{\partial y} \end{bmatrix} = [G]\{p\}^e \qquad (5.82)$$

The volumetric strain for plane strain conditions is given by

$$\varepsilon_x + \varepsilon_y = [B_\Delta]\{\delta\}^e \qquad (5.83)$$

The effective stress within an element may be written as

$$\{\sigma\} = [D]\{\varepsilon\} + \{\sigma_0\} \qquad (5.84)$$

where $[D]$ is the elasticity matrix and $\{\sigma_0\}$ is the initial stress vector.
The fluid flux at any point is given by

$$\{q\} = [H][G]\{p\} + [H]\{\rho_2 F\} \qquad (5.85)$$

where $[H]$ is the permeability matrix (2×2) and $\{\rho_2 F\}$ is the fluid body force vector.

Expressing the volume integrals in equation 5.78 as the sum of the integrals over each element and introducing the above expressions for displacements, pore pressures and effective stresses, the functional given by equation 5.78 can be written as

$$\Omega_t(u, p) = \tfrac{1}{2}\left[(\{\delta\}^e)^T[K_1]\{\delta\}^e + (\{\delta\}^e)^T\{M_1\}*g\right.$$
$$-(\{\delta\}^e)^T\{M_2\}*g + (\{\delta\}^e)^T[c]\{p\} - \tfrac{1}{2}\left[g*\{p\}^T[K_2]\{p\}\right]$$
$$\left. -\tfrac{1}{2}\left[g*\{p\}^t\{M_3\}*g\right] - (\{\delta\}^e)^T\{P_1\} + g*\{p\}^T\{P_2\} \right. \qquad (5.86)$$

in which

$$[K_1] = \sum_{m=1}^{M} \int_{V_m} [B_u]^T[D][B_u]dV_m$$

$$[K_2] = \sum_{m=1}^{M} \int_{V_m} [G]^T[H][G]dV_m$$

$$[C] = \sum_{m=1}^{M} \int_{V_m} [B_\Delta]^T\{N_p\}dV_m$$

$$\{M_1\} = \sum_{m=1}^{M} \int_{V_m} [B_u]^T\{\sigma_0\}dV_m$$

$$\{M_2\} = \sum_{m=1}^{M} \int_{V_m} [N_u]^T\{\rho F\}dV_m$$

$$\{M_3\} = \sum_{m=1}^{M} \int_{V_m} [G]^T[H]\{\rho_2 F\}dV_m$$

$$\{P_1\} = \sum_{m=1}^{M} \int_{S1_m} [N_u][N_u]^T\{T\}dS_m$$

$$\{P_2\} = \sum_{m=1}^{M} \int_{S2_m} [N_p][N_p]^T\{Q\}dS_m$$

where M is the number of elements and m is the element number.

By applying the variational principle and minimising the functional given in equation 5.86 with respect to the nodal values, the following set of matrix equations can be obtained.

$$[K_1]\{\delta\}^e + [C]\{p\} = -\{M_1\} + \{M_2\} + \{P_1\} \tag{5.87}$$

$$[C]^T\{\delta\}^e - g^*[K_2]\{p\} = g^*\{M_3\} - g^*\{P_2\} \tag{5.88}$$

These two equations have the following physical meanings. Equation 5.87 is the equilibrium equation in which $[K_1]\{\delta\}^e$ is the nodal force vector due to the straining of soil skeleton, $[C]\{p\}$ is the nodal force vector resulting from the pore pressure, $\{M_1\}$ is the load vector due to initial stresses, $\{M_2\}$ is the load vector due to the body forces and $\{P_1\}$ is the specified boundary traction vector. Locally applied loads may be included in the vector $\{P_1\}$. Equation 5.88 represents the flow equation in which $[C]^T\{\delta\}^e$ is the volumetric strain, $g^*[K_2]\{p\}$ is the fluid inflow due to pore pressure, $g^*\{M_3\}$ is due to gravity forces and $g^*\{P_2\}$ is due to specified boundary flow. Local drainages applied at nodes may be added to the vector $\{P_2\}$.

The time domain can be divided into a number of small time intervals and for each time increment the following equations are solved on the assumption that displacements and pore pressures have been already determined for the previous time interval.

$$\begin{bmatrix} [K_1] & [C] \\ [C]^T & -[K_2] \end{bmatrix} \begin{bmatrix} \{\delta\}^e \\ \{p\} \end{bmatrix} = \begin{bmatrix} \{F_1\} \\ \{F_2\} \end{bmatrix} \tag{5.89}$$

where
$$\{F_1\} = -\{M_1\} + \{M_2\} + \{P_1\}$$
$$\{F_2\} = g^*\{M_3\} - g^*\{P_2\}$$

5.6.3 Application: Strip load on elastic layer. Plane strain

The procedure described for the analysis of consolidation problems has been applied by Hwang et al. (1971) to determine the settlement of a porous elastic layer of finite thickness resting on a smooth impervious base. The same problem has also been analysed by Gibson et al. (1970) and hence a comparison of the results obtained by the two methods is given in *Figure 5.11* and *Figure 5.24*. In *Figure 5.11* the variation of the rate of consolidation with the adjusted time factor for different depths at the centre line of the loaded layer is shown. The variation of settlement with time beneath the mid-point of the layer for various ratios of Poisson's ratio is given in *Figure 5.24*. From these two figures, the excellent agreement between the finite element solution and the closed form solution can be observed.

5.7 SETTLEMENT OF FOOTINGS—NON-LINEAR ANALYSIS

5.7.1 Introduction

Engineers have long been able to estimate the settlement of footings on the supporting soil using the theory of elasticity. To estimate the ultimate bearing capacity it is necessary to use the theory of plasticity. Several solutions based on the theory of plasticity exist only for simplified problems which include the assumptions of isotropic material properties, homogeneous nature of the supporting soil mass and ideal or rigid-plastic behaviour. Thus the complexities of the non-linear analysis have tended to create a 'non-linear barrier' which has prevented engineers from gaining an appreciation of the load-settlement characteristics based on a consideration of the real behaviour of the soil.

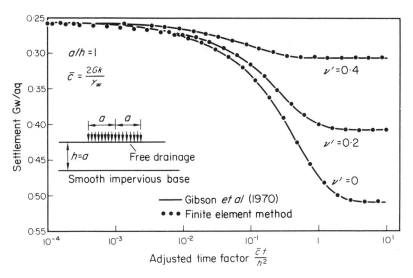

Figure 5.24   Rate of settlement for a strip load (after Hwang et al. (1971). Courtesy Canadian Geot. Jour.)

However, with the advent of high-speed computers and the development of simpler numerical techniques such as the finite element method, tools are now available for predicting settlement behaviour based on the realistic stress-strain characteristics. The finite element method has been used to a great extent in the linear elastic analysis of soil mechanics problems and to a less extent in the non-linear analysis. In this section, a computational process using the finite element method for the elasto-plastic analysis of soil mechanics problems is described and the application of this technique to the particular problem of estimating the settlement of strip footings in a soil mass is discussed.

5.7.2   Elasto-plastic analysis

For any method of elasto-plastic analysis, the three basic requirements are

(1) Linear (elastic) constitutive relationship between stresses and strains.
(2) Yield criterion establishing the onset of plastic flow.
(3) Non-linear (plastic) constitutive relationship between stresses and strains.

Linear constitutive relations.   The stresses $\{\sigma\}$ and the strains $\{\varepsilon\}$, within the elastic range, can be related in terms of Hookean elasticity as

$$\{\sigma\} = [D]\{\varepsilon\} \tag{5.90}$$

Where $[D]$ is the elasticity matrix consisting terms based on modulus of elasticity, E and Poisson's ratio, v. The elasticity matrix can be formed either for isotropic or anisotropic material properties and depending on whether the problem is plane stress, plane strain, axisymmetric or three-dimensional. Separate expressions for these various cases have been given by Timoshenko (1969) and Zienkiewicz (1971).

Yield criterion.   A yield criterion is a hypothesis used in the multiaxial states of stress, to provide a functional relation among the stresses at the onset of yield.

Isotropic.   For problems in soil mechanics several yield criteria have been proposed by Bishop (1966). Von Mises yield criterion has been widely used for metals whereas Coulomb's criterion has been generally adopted for soils. In the case of soils, it is essential to use a yield criterion which consists of terms corresponding to the cohesion and the friction of the material. Such a yield criterion has been proposed by Drucker (1952). Drucker's yield criterion is a modified version of von Mises yield criterion including Coulomb hypothesis and is of the form

$$\alpha J_1 + J_2^{\frac{1}{2}} = K \tag{5.91}$$

where $J_1 = \sigma_x + \sigma_y + \sigma_z;$

$$J_2 = \tfrac{1}{6}[(\sigma_x - \sigma_y)^2 + (\sigma_y - \sigma_z)^2 + (\sigma_z - \sigma_x)^2] + \tau_{xy}^2 + \tau_{yz}^2 + \tau_{zx}^2$$

α and K are constants depending on cohesion and friction of the material. The constants α and K are related to cohesion, c and angle of friction, ϕ as

$$\alpha = \frac{\tan \phi}{\sqrt{(9 + 12 \tan^2 \phi)}} \tag{5.92a}$$

$$K = \frac{3c}{\sqrt{(9 + 12 \tan^2 \phi)}} \tag{5.92b}$$

Thus the Drucker's yield criterion for two-dimensional problems with isotropic material properties can be written as

$$F = \alpha(\sigma_x + \sigma_y + \sigma_z) + \{\tfrac{1}{6}[(\sigma_x - \sigma_y)^2 + (\sigma_y - \sigma_z)^2 + (\sigma_z - \sigma_x)^2] + \tau_{xy}^2\}^{\frac{1}{2}} = K \tag{5.93}$$

Anisotropic.   For initially anisotropic materials in geomechanics, Hill's (1950) yield criterion has been extended by Pariseau (1968). Pariseau's criterion is similar to Drucker's criterion and is of the form

$$F = [\alpha_{12}(\sigma_x - \sigma_y)^2 + \alpha_{23}(\sigma_y - \sigma_z)^2 + \alpha_{31}(\sigma_z - \sigma_x)^2 + \alpha_{44}\tau_{xy}^2]^{\frac{1}{2}}$$
$$+ (\alpha_{11}\sigma_x + \sigma_{22}\sigma_y + \alpha_{33}\sigma_z) = K \tag{5.94}$$

where $K = 1$ and α's are material constants defined in terms of unconfined tensile and compressive strengths in the anisotropic reference axes.

Non-linear constitutive relations.   The associated flow rules of the incremental theory of plasticity specify the normality principle to relate the increments of strains to the increments of stress. The normality principle states that the plastic strain vector for a particular point on the yield surface is normal to the yield surface at that point.

Thus,
$$\{d\varepsilon^p\} = \lambda . \left\{\frac{\partial F}{\partial \sigma}\right\}$$
(5.95)

where $\{d\varepsilon^p\}$ is the plastic strain vector and λ is the proportionality constant.

Total stress-strain relations.   During an increment of load, the total strain in the plastic region can be considered to consist of elastic and plastic strains.

$$\{d\varepsilon\} = \{d\varepsilon^e\} + \{d\varepsilon^p\}$$
(5.96)

The elastic strain increments are related to the stress increments by the elasticity matrix $[D]$ as

$$\{d\varepsilon^e\} = [D]^{-1}\{d\sigma\}$$
(5.97)

Thus the total incremental strain is

$$\{d\varepsilon\} = [D]^{-1}\{d\sigma\} + \lambda \left\{\frac{\partial F}{\partial \sigma}\right\}$$
(5.98)

Differentiating the yield criterion, whether it is for isotropic or anisotropic,

$$\frac{\partial F}{\partial K}dK = \frac{\partial F}{\partial \sigma_x}d\sigma_x + \frac{\partial F}{\partial \sigma_y}d\sigma_y + \frac{\partial F}{\partial \sigma_z}d\sigma_z + \frac{\partial F}{\partial \tau_{xy}}d\tau_{xy}$$
(5.99)

Letting
$$A = \frac{\partial F}{\partial K}dK . \frac{1}{\lambda}$$

Equation 5.99 can be rewritten as

$$\frac{\partial F}{\partial \sigma_x}d\sigma_x + \frac{\partial F}{\partial \sigma_y}d\sigma_y + \frac{\partial F}{\partial \sigma_z}d\sigma_z + \frac{\partial F}{\partial \tau_{xy}}d\tau_{xy} - A\lambda = 0$$
(5.100)

Thus the relationship between total stresses and strains during an increment can be written in matrix form as

$$
\begin{bmatrix} d\varepsilon_x \\ d\varepsilon_y \\ d\varepsilon_z \\ d\gamma_{xy} \\ 0 \end{bmatrix} = \begin{bmatrix} |D|^{-1} & \begin{matrix} \frac{\partial F}{\partial \sigma_x} \\ \frac{\partial F}{\partial \sigma_y} \\ \frac{\partial F}{\partial \sigma_z} \\ \frac{\partial F}{\partial \tau_{xy}} \end{matrix} \\ \frac{\partial F}{\partial \sigma_x} \frac{\partial F}{\partial \sigma_y} \frac{\partial F}{\partial \sigma_z} \frac{\partial F}{\partial \tau_{xy}} & -A \end{bmatrix} \begin{bmatrix} d\sigma_x \\ d\sigma_y \\ d\sigma_z \\ d\tau_{xy} \\ \lambda \end{bmatrix}
$$
(5.101)

The parameter 'A' is obviously zero for perfectly plastic materials. On the other hand, for work-hardening materials, the value of 'A' has been shown to be the slope of the work-hardening curve by Zienkiewicz *et al.* (1969).

Considering the soils to behave as perfectly plastic material, the value of 'A' is zero. Therefore the relationship in the form given by equation 5.101 is not suitable for the purposes of computation. However, the stress increments can be conveniently expressed in terms of the strain increments.

If equation 5.101 is written as

$$
\begin{bmatrix} \{d\varepsilon\} \\ 0 \end{bmatrix} = \begin{bmatrix} [D]^{-1} & \left\{\dfrac{\partial F}{\partial \sigma}\right\} \\ \left\{\dfrac{\partial F}{\partial \sigma}\right\}^{T} & -A \end{bmatrix} \begin{bmatrix} \{d\sigma\} \\ \lambda \end{bmatrix} \tag{5.102}
$$

$$
\{d\varepsilon\} = [D]^{-1}\{d\sigma\} + \lambda \left\{\frac{\partial F}{\partial \sigma}\right\} \tag{5.103}
$$

$$
0 = \left\{\frac{\partial F}{\partial \sigma}\right\}^{T} \{d\sigma\} - A\lambda \tag{5.104}
$$

Solving these two equations, λ and $\{d\sigma\}$ can be obtained.

$$
\lambda = \left[A + \left\{\frac{\partial F}{\partial \sigma}\right\}^{T} [D] \left\{\frac{\partial F}{\partial \sigma}\right\} \right]^{-1} \left\{\frac{\partial F}{\partial \sigma}\right\}^{T} [D]\{d\varepsilon\} \tag{5.105}
$$

$$
\{d\sigma\} = [D]\{d\varepsilon\} - [D]\left\{\frac{\partial F}{\partial \sigma}\right\}
$$
$$
\times \left[A + \left\{\frac{\partial F}{\partial \sigma}\right\}^{T} [D] \left\{\frac{\partial F}{\partial \sigma}\right\} \right]^{-1} \left\{\frac{\partial F}{\partial \sigma}\right\}^{T} [D]\{d\varepsilon\} \tag{5.106}
$$

or
$$
\{d\sigma\} = [Dep]\{d\varepsilon\} \tag{5.107}
$$

where $[Dep]$ is the elasto-plastic matrix given by

$$
[Dep] = [D] - [D]\left\{\frac{\partial F}{\partial \sigma}\right\}\left[A + \left\{\frac{\partial F}{\partial \sigma}\right\}^{T} [D]\left\{\frac{\partial F}{\partial \sigma}\right\} \right]^{-1} \left\{\frac{\partial F}{\partial \sigma}\right\}^{T} [D] \tag{5.108}
$$

5.7.3 Finite element method—initial stress technique

The details of the finite element method are described in several papers and books but the notations used in this section are similar to those given by Zienkiewicz (1971). For the non-linear analysis, first a linear elastic analysis of the problem of known configuration and boundary conditions has to be performed. In the displacement method, given a set of loads, the appropriate displacements $\{\delta\}$ at the nodes are determined throughout the continuum. From the displacements, the strains are determined as

$$
\{\varepsilon\} = [B]\{\delta\} \tag{5.109}
$$

Where the matrix $[B]$ defines the relation between strains and deformations on the basis of small deformation theory.

From the strains, the stresses are obtained using the constitutive relationship for the material. In the case of linear elasticity,

$$
\{\sigma\} = [D]\{\varepsilon\} \tag{5.110}
$$

where $[D]$ is the elasticity matrix.

At the onset of plastic flow, the stresses are related to the strains as given in equation 5.101.

$${\sigma} = [\text{Dep}] {\varepsilon} \tag{5.111}$$

where $[\text{Dep}]$ is the elasto-plastic matrix.

In the finite element analysis, equation 5.110 is used for an element when the stresses in that element are in the elastic range whereas for the same element equation 5.111 should be used when the material behaviour is elasto-plastic. It can be seen immediately that for all the yielded elements, this involves changing the original stiffness matrix and assembling the structure stiffness matrix again and then solving a new system of simultaneous equations. It is apparent that this process will be inefficient because of the time involved in repeatedly solving different elastic problems. Therefore Zienkiewicz et al. (1969) proposed an 'initial stress technique' which uses the original elasticity matrix repeatedly and estimates the corrections in the stresses in order to reproduce the appropriate constitutive law.

In the initial stress approach, the solution of a non-linear problem is obtained through piecewise linearisation. During an increment of load, the problem is solved on the basis of linear elasticity using the original $[D]$ matrix.

For an element the stresses are obtained as

$${\sigma_1} = [D]{\varepsilon_1} \tag{5.112}$$

If the element is in the elastic range these stresses are correct. On the other hand if the element has yielded the stresses obtained using equation 5.112 are not valid. The actual stresses should have been obtained using the elasto-plastic matrix as

$${\sigma_1'} = [\text{Dep}]{\varepsilon_1} \tag{5.113}$$

The difference between the stresses ${\sigma_1}$ and ${\sigma_1'}$ is termed the 'initial stress'.

$${\sigma_1''} = {\sigma_1} - {\sigma_1'} \tag{5.114}$$

This difference has a definite physical meaning. The initial stress actually represents the incapability of the material to sustain stresses which are in excess of its actual strength. Thus for the structure, to withstand the additional loads resulting from the initial stresses, the excess stresses should be transferred to other parts of the structure during the subsequent steps. This is accomplished by converting these initial stresses into body forces as follows:

$${P_1^e} = \int [B]^T {\sigma_1''} \, d(\text{Vol.}) \tag{5.115}$$

During the next stage of computation, this system of body forces is applied to the structure with the original elasticity matrix and the structure is allowed to deform further. This produces an additional set of strains and corresponding stresses. Again, it will be found that the new stresses exceed those allowable by the non-linear constitutive relation. Thus it may be necessary to redistribute these initial stresses again. The process is repeated until the body forces resulting from the initial stresses become negligible. When the process finally converges, the non-linear constitutive law will be satisfied.

5.7.4 Application: Strip footing on layered soil

The application of the finite element method using the initial stress technique, to the elasto-plastic analysis of soil mechanics problems is presented in this section. The behaviour of strip footing on soil is discussed with reference to two specific problems—one with the footing resting on single layered clay and the other with the footing on a two-layered clay system.

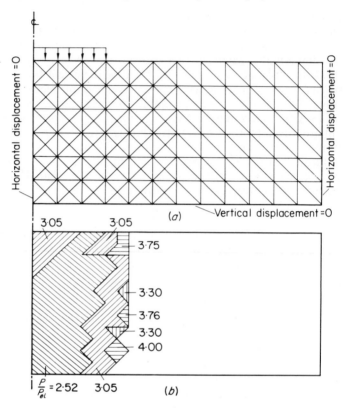

Figure 5.25(a) Strip footing on a single layer. Finite element mesh

(b) Strip footing on a single layer. Growth of plastic zone

The settlement of the strip footing on a single layer of clay has been predicted by Hoeg *et al.* (1968) using the lumped parameter model and by Radhakrishnan and Reese (1969) using the finite element method. The results given in this section are for the same problem as that analysed by Hoeg *et al.* (1968). The strip footing was assumed to be resting on a single clay layer with the following soil properties—modulus of elasticity of 206 700 kN . m^{-2}. Poisson's ratio of 0.3 and constant cohesive strength of 120.6 kN . m^{-2}. The configuration of the finite element mesh is shown in *Figure 5.25a* whereas *Figure 5.25b* shows the spread of plastic zones for various ratios of P/P_{el} where P_{el} is the load required to initiate yielding. In this analysis, the footing was considered to be rigid and instead of applying the strip load in increments, uniform vertical rigid displacements were applied in increments at all the nodal points within the width of the footing. *Figure 5.26* shows the plot of the pressure-settlement curve for the footing and a comparison is made with the results of Hoeg *et al*. The failure load of $606 \, \mathrm{kN.m^{-2}}$ predicted by the initial stress method compares favourably with the Prandtl's failure load of $620 \, \mathrm{kN.m^{-2}}$.

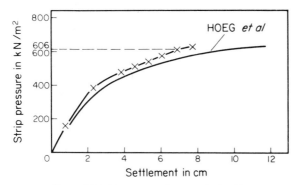

Figure 5.26   *Pressure-settlement relationship*

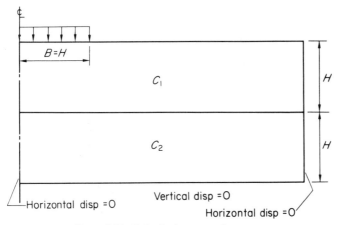

Figure 5.27   *Strip footing on two layers*

The second problem illustrates the behaviour of strip footing in a two-layered clay system, shown in *Figure 5.27*. Again uniform rigid vertical displacements were applied in increments over the width of the footing in order to determine the strip pressure. Two different values of the ratios of the cohesive strengths of the two layers were considered for the analysis. *Figure 5.28* shows the pressure-settlement characteristics for the case of the clay layers with $c_1/c_2 = 0.5$ whereas the same has been plotted for the case of $c_1/c_2 = 0.25$ in *Figure 5.29*. The ultimate bearing capacity in this case depends on the relative cohesive strengths of layers as well as the ratio of the depth of the top layer (H) to the half-width of the footing (B). Radhaknishnan and Reese (1969) have observed that for a ratio of H/B greater than 1.3, the lower layer will have no effect on the ultimate bearing capacity. In this example, since the ratio H/B is equal to 1.0, the lower layer will have some effect on the ultimate bearing capacity due to its higher value of cohesive strength. This effect can be observed from *Figure 5.28* and *Figure 5.29*. The finite element solution predicts an ultimate bearing capacity of 313 kN.m^{-2} whereas using Prandtl's solution for the upper layer it is 310 kN.m^{-2}, for the case of $c_1/c_2 = 0.5$. Similarly the finite element solution predicts a value 171 kN.m^{-2} for the ultimate bearing capacity in the case of $c_1/c_2 = 0.25$, whereas it is 155 kN.m^{-2} using Prandtl's solution.

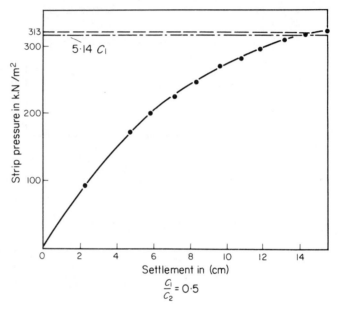

Figure 5.28　Pressure-settlement relationship $c_1/c_2 = 0.5$

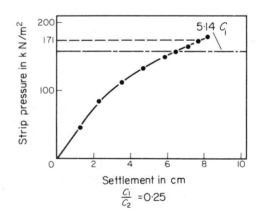

Figure 5.29　Pressure-settlement relationship $c_1/c_2 = 0.25$

From the two examples illustrated it can be observed that the results obtained by the finite element method are satisfactory and encouraging. The method described herein can be successfully extended to the multi-layered systems that are difficult to analyse by conventional methods.

BIBLIOGRAPHY

ABOSHI, H., YOSHIKUNI, H., and MARUYAMA, S. (1970). 'Constant Loading Rate Consolidation Test', *Soils and Foundations*, **10**, 1, 44–56.

BAKER, A. L. L. (1957). *Raft Foundations*, 3rd ed., Concrete Publications Ltd., London.

BIOT, M. A. (1941a). 'General Theory of Three-Dimensional Consolidation', *Jour. Appl. Phys.*, **12**, 155–164.

BIOT, M. A. (1941b). 'Consolidation Settlement under a Rectangular Load Distribution', *Jour. Appl. Phys.*, **12**, 426–430.

BIOT, M. A. (1955). 'Theory of Elasticity and Consolidation for a Porous Anisotropic Solid', *Jour. Appl. Phys.*, **12**, 182–185.

BIOT, M. A. (1956a). 'Theory of Deformation of a Porous Viscoelastic Anisotropic Solid', *Jour. Appl. Phys.*, **27**, 459–467.

BIOT, M. A. (1956b). 'General Solutions of the Equations of Elasticity and Consolidation for a Porous Material', *Jour. Appl. Mechs., Trans. ASME*, **78**, 91–96.

BIOT, M. A. (1963). 'Theory of Stability and Consolidation of a Porous Medium under Initial Stress', *Jour. Math. Mech.*, **12**, 521–541.

BIOT, M. A., and CLINGAN, F. M. (1941). 'Consolidation Settlement of a Soil with an Impervious Top Surface', *Jour. Appl. Phys.*, **12**, 578–581.

BIOT, M. A., and CLINGAN, F. M. (1942). 'Bending Settlement of a Slab Resting on a Consolidating Foundation', *Jour. Appl. Phys.*, **13**, 35–40.

BIOT, M. A., and WILLIS, D. G. (1957). 'The Elastic Coefficients of the Theory of Consolidation', *Jour. Appl. Mech., Trans. ASME*, **24**, 594–601.

BISHOP, A. W. (1966). 'The Strength of Soils as Engineering Materials', *Geotechnique*, **16**, 91–128.

CHRISTIAN, J. T., and BOEHMER, J. W. (1970). 'Plane Strain Consolidation by Finite Elements', *Jour. Soil Mech. Fdn. Div., ASCE*, **96**, 1435–1457.

CRYER, C. W. (1963). 'A Comparison of the Three-Dimensional Theories of Biot and Terzaghi', *Quart. Jour. Mech. Appl. Math.*, **16**, 401–412.

DAVIS, E. H., and RAYMOND, G. (1965). 'A Non-Linear Theory of Consolidation', *Geotechnique*, **15**, 2, 161–173.

DAVIS, E. H., and LEE, I. K. (1969). 'One-Dimensional Consolidation of Layered Soils', *Proc. 7th Int. Conf. Soil Mechs. and Fdn. Eng.*, **2**, 65–72.

DAVIS, E. H., and POULOS, H. G. (1972). 'Rate of Settlement under Two- and Three-Dimensional Conditions', *Geotechnique*, **22**, 1, 95–114.

D'APPOLONIA, D. J., and LAMBE, T. W. (1970). 'Method for Predicting Initial Settlement', *Jour. Soil Mech. Fdn. Eng. Div., ASCE*, **96**, SM 2, 523–544.

DE JONG, G. J. (1957). 'Application of Stress Functions to Consolidation Problems', *Proc. 4th Int. Conf. Soil Mechs. and Fdn. Eng.*, **1**, 320–323.

DRUCKER, D. C., and PRAGER, W. (1952). 'Soil Mechanics and Plastic Analysis of Limit Design', *Quarterly of Appl. Maths.*, **10**, 2, 157–165.

GIBSON, R. E., and MCNAMEE, J. (1957). 'The Consolidation Settlement of a Load Uniformly Distributed over a Rectangular Area', *Proc. 4th Int. Conf. Soil Mech. and Fdn. Eng.*, **1**, 297–299.

GIBSON, R. E., ENGLAND, G. L., and HUSSEY, M. J. L. (1967). 'The Theory of One-Dimensional Consolidation of Saturated Clays'. I. 'Finite Non-Linear Consolidation of Thin Homogeneous Layers', *Geotechnique*, **17**, 261–273.

GIBSON, R. E., SCHIFFMAN, R. I., and PU, S. L. (1970). 'Plane Strain and Axially Symmetric Consolidation of a Clay Layer on a Smooth Impervious Base', *Quart. Jour. Mech. Appl. Maths.*, **23**, 4, 505–520.

GORBUNOV-POSSADOV, M. J. (1959). *Tables for the Design of Thin Slabs on Elastic Subsoil*, Moscow, 65–96.

GRAY, H. (1945). 'Simultaneous Consolidation of Contiguous Layers of Unlike Compressible Soils', *Trans. ASCE*, **110**, 1327–1344.

HAIN, S. J., and LEE, I. K. (1973). 'A Rational Analysis of Raft Foundations', *UNICIV Report R–102*, University of New South Wales, Australia.

HILL, R. (1950). *The Mathematical Theory of Plasticity*, Clarendon Press, Oxford.

HOEG, K., CHRISTIAN, J. T., and WHITMAN, R. V. (1968). 'Settlement of Strip Load on Elasto-Plastic Soil', *Jour. Soil Mech. and Fdn. Div., ASCE*, **94**, SM 2, 431–445.

HOGG, A. H. A. (1938). 'Equilibrium of a Thin Plate, Symmetrically Loaded, Resting on an Elastic Foundation of Infinite Depth', *Phil. Mag.*, **25**, 576–582.

HOGG, A. H. A. (1944). 'Equilibrium of a Thin Slab on an Elastic Foundation of Finite Depth', *Phil. Mag.*, **35**, 265–276.

HOLL, D. L. (1938). 'Thin Plates on Elastic Foundations', *5th Congress Appl. Mech.*, 71–74, Cambridge, Mass.

HWANG, C. T., MORGENSTERN, N. R., and MURRAY, D. W. (1971). 'On Solutions of Plain Strain Consolidation Problems by Finite Element Methods', *Canadian Geot. Jour.*, **8**, 1, 109–118.

JANBU, N. (1965). 'Consolidation of Clay Layers Based on Non-Linear Stress/Strain', *Proc. 6th Int. Conf. Soil Mech. and Fdn. Eng.*, **2**, 83–87.

LEE, I. K. (1968). *Soil Mechanics. Selected Topics*, Butterworths, London.

LEE, I. K., and HARRISON, H. (1970). 'Structure and Foundation Interaction Theory', *Jour. Struct. Div.*, ASCE, **96**, ST 2, 177–197.

LEE, I. K., and BROWN, P. T. (1972). 'A Theoretical Analysis of the Interaction Between a Structure and a Continuous Foundation', *UNICIV Report R–80*. University of New South Wales, Australia.

LOWE, J., JONAS, E., and OBRCIAN, V. (1969). 'Controlled Gradient Consolidation Test', *Jour. Soil Mech. and Fdns. Div.*, ASCE, **95**, SM 1, 77–97.

MANDEL, J. (1957). 'Consolidation des couches d'argiles', *Proc. 4th Int. Conf. Soil Mech. and Fdn. Eng.*, **1**, 360–367.

MANDEL, J. (1961). 'Tassements produits par la consolidation d'une couche d'argile de grande épaisseur', *Proc. 5th Int. Conf. Soil Mech. and Fdn. Eng.*, **1**, 733–736.

MCNAMEE, J., and GIBSON, R. E. (1960). 'Plane Strain and Axially Symmetric Problems of the Consolidation of a Semi-Infinite Clay Stratum', *Quart. Jour. Mech. and Appl. Math.*, **13**, 2, 210–227.

MURPHY, G. (1937). 'Stresses and Deflections in Loaded Rectangular Plates on Elastic Foundations', *Iowa State College Bulletin No. 135*.

PARISEAU, N. G. (1968). 'Plasticity Theory for Anisotropic Rocks and Soils', *10th Symposium on Rock Mechanics*, Austin, Texas.

PICKETT, G., JONES, W. C., and MCCORMICK. (1951). 'Deflections, Moments, and Reactive Pressures for Concrete Pavements', *Eng. Exp. Station*, Kansas State College. Manhattan, Kansas.

PRZEMIENIECKI, J. S. (1968). *Theory of Matrix Structural Analysis*, McGraw-Hill, New York.

RADHAKRISHNAN, N., and REESE, L. C. (1969). 'Behaviour of Strip Footings on Layered Cohesive Soils', *Proc. Symp. on Application of Finite Element Methods in Civil Engineering*, Vanderbilt University, Tennessee.

ROWE, P. W., and SHIELDS, D. H. (1965). 'The Measured Horizontal Coefficient of Consolidation of Laminated, Layered or Varied Clays', *Proc. 6th Int. Conf. Soil Mech. and Fdns. Eng.*, **1**, 342–344.

SANDHU, R. S., and WILSON, E. L. (1969). 'Finite Element Analysis of Seepage in Elastic Media', *Jour. Eng. Mechs. Div.*, ASCE, **95**, 641–652.

SCHIFFMAN, R. L. (1958). 'Consolidation of Soil under Time-Dependent Loading and Varying Permeability', *Proc. HRB*, **37**, 584–617.

SCHIFFMAN, R. L., CHEN, A. T-F., and JORDAN, J. C. (1969). 'An Analysis of Consolidation Theories', *Jour. Soil Mech. and Fdns. Div.*, ASCE, **95**, SM 1, 285–312.

SEVERN, R. T. (1966). 'The Solution of Foundation Mat Problems by Finite Element Methods', *Struct. Eng.*, **44**, 6, 223–228.

SMITH, R. E., and WAHLS, H. E. (1969). 'Consolidation under Constant Rate of Strain', *Jour. Soil Mech. and Fdns. Div.*, ASCE, **95**, SM 2, 519–539.

TERZAGHI, K. V. (1943). *Theoretical Soil Mechanics*, John Wiley & Sons, New York.

TIMOSHENKO, S., and GOODIER, J. N. (1969). *Theory of Elasticity*, McGraw-Hill, New York.

WESTERGAARD, H. M. (1947). 'New Formulas for Stresses in Concrete Pavements of Airfields', *Proc. ASCE*, **73**, 5, 687–701.

WIZZA, A. E., CHRISTIAN, J. T., and DAVIS, E. H. (1971). 'Consolidation at a Constant Rate of Strain', *University of Sydney Research Report R172*.

ZIENKIEWICZ, O. C. (1971). *The Finite Element Method in Engineering Science*, McGraw-Hill, London.

ZIENKIEWICZ, O. C., and CHEUNG, Y. K. (1965). 'Plates and Tanks on Elastic Foundations—An Application of Finite Element Method', *Int. Jour. Solids and Structures*, **1**, 451–461.

ZIENKIEWICZ, O. C., and CHEUNG, Y. K. (1969). *Finite Element Method in Structural and Continuum Mechanics*, McGraw-Hill, London.

ZIENKIEWICZ, O. C., VALLIAPPAN, S., and KING, I. P. (1969). 'Elasto-Plastic Solutions of Engineering Problems Initial Stress, Finite Element Approach', *Int. Jour. Num. Meth. Eng.*, **1**, 75–100.

Chapter 6

Stability and Earth Pressures

I. K. Lee and J. R. Herington

6.1 INTRODUCTION

Traditional methods of investigating the stability of foundations, slopes and retaining structures, invoke sufficient assumptions until the analysis can be completed by recourse to statics and a failure theory. The basic problem of predicting deformations up to failure of the soil mass is an exceedingly difficult one to solve satisfactorily. In recent years significant progress has been made in the determination of the stress-strain properties of soils, and the application of these constitutive laws to the prediction of the performance of structures.

There is an interesting contrast between the history of the development of these analyses of soil masses and the analyses of steel structures. Simple constitutive laws of the latter material were established prior to the general usage of the indeterminate analyses and the need for a stability or limit analysis was not paramount. By contrast, the complexity of the stress-strain laws of soils led to the use of stability analyses at an early stage in the history of soil mechanics.

When the stability analysis is applied to a retaining structure it is traditional to assume a failure surface in order to satisfy the elementary kinematics of the problem. Since the rigid-plastic soil model is implied an infinitesimal movement along the failure surface is sufficient to develop the full shear strength, conveniently expressed by the c, ϕ parameters. Then to render the analysis statically determinate it is also necessary to assume a stress state along the soil-wall interface. An infinitesimal relative movement along the interface is sufficient to develop either the 'rough' (maximum interfacial shear stress) positive or negative interfacial state. Conversely, however, zero relative movement can be associated with any stress state intermediate between the two limiting rough states. By comparison, a real soil requires a finite relative movement to develop the intermediate stress state.

When reviewing the state of art of the stability analyses of soil structures it is necessary to differentiate between

(1) the accuracy of the 'approximate' static analyses relative to the correct solutions for a rigid-plastic material; and
(2) the applicability of the rigid-plastic model to the soil.

Unless the evaluation is carried out in these two stages it is extremely difficult to make generalised conclusions regarding the applicability of a particular traditional analysis. The accuracy of the existing approximate solutions, and the situation for which any particular analysis is applicable, can be established by comparing predicted values with the corresponding values obtained by the plasticity theory for a rigid plastic, $c-\phi-\gamma$, rigid plastic material, obeying either the associated or non-associated flow rule material. The applicability of the 'correct' solution for this idealised material can then be examined by comparing predicted values with the experimental data now available.

Use of the plasticity theory leads to a consideration of the type of wall movement thus Terzaghi's 'fundamental fallacy in earth pressure computations' can be resolved using this approach.

This chapter includes a detailed discussion of the techniques used to establish upper and lower bound solutions and is written to extend the contribution in the earlier volume (Davis (1968)). Solutions for retaining walls are presented and the applicability of these solutions is established.

6.2   STRESS AND VELOCITY FIELDS. NUMERICAL TECHNIQUES

6.2.1   Stress Fields

The conditions within a two-dimensional plastic field are established by combining the equilibrium equations with the Mohr-Coulomb failure criterion. This leads to the hyperbolic equations which define two families of curves in the x–y plane. These curves are surfaces along which shear failure criterion is satisfied and, for convenience, these two families of curves are referred to as α lines and β lines or the stress characteristics. The complete system forms a stress field, and an analysis of this field leads to a lower bound solution for the specific problem.

The β lines are characterised by the fact that their slope is given by the first solution of the hyperbolic equations, and the α lines are characterised by the second solution of the equations.

It was shown that (Sokolovski (1956))

$$\frac{dy}{dx} = \tan(\theta \pm (\tfrac{1}{4}\pi - \tfrac{1}{2}\phi))$$

$$= \tan(\theta \pm \mu) \tag{6.1}$$

for the β and α lines, respectively. θ is the anticlockwise angle from the vertical (x) axis to the major principal stress direction. The two families intersect at an angle of 2μ and are orientated at an angle of $\pm\mu$ to the major principal stress direction.

The change in mean normal stress,† along the stress characteristics, $d\sigma_p$, is given by the expressions

$$d\sigma_p \pm 2\sigma_p \cdot \cot 2\mu \cdot d\theta = \mp \frac{\gamma \sin(\theta \mp \mu)}{\sin 2\mu \cdot \cos(\theta \pm \mu)} \cdot dx \tag{6.2a}$$

$$= a(\theta) \cdot dx \quad \text{or} \quad b(\theta) \cdot dx \tag{6.2b}$$

† $\sigma_p = \dfrac{\sigma_1 + \sigma_3}{2} + c \cot \phi$

The subscript p refers to a 'programme' stress defined by

$$\sigma_{xp} = \sigma_x + c \cot \phi$$
$$\sigma_{yp} = \sigma_y + c \cot \phi$$

where σ_x, σ_y are the real stresses. This technique allows a $c-\phi-\gamma$ material to be analysed as a $\phi-\gamma$ material.

When suitable stress boundary conditions can be recognised equations 6.1 and 6.2 enable the stress field and the associated stress increments to be determined. There are three basic types of stress field which can be involved in a numerical analysis.

(1) Cauchy *(Figure 6.1)*.
(2) Goursat *(Figures 6.2, 6.3)*.
(3) Mixed Boundary *(Figure 6.4)*.

Consider a stress-free surface as shown in *Figure 6.1* which illustrates the Cauchy problem. The major principal stress can be either parallel to or normal to the surface and initially let us consider the case of passive failure so that σ_1 is parallel to the surface. For any surface point the value of σ is $\frac{1}{2}\sigma_1$ and θ is $(\frac{1}{2}\pi - i)$ recalling that θ is considered positive anticlockwise.

Select two points such as A and B. Then it is seen that the α line from A will intersect the β line from B at point C within the soil mass. As shown by Prager and Hodge (1951) for a $\phi = 0$, weightless material, and by Sokolovski (1956) for a $c-\phi-\gamma$ material, the location of C, hence the value of σ_C, θ_C, can be obtained from the finite difference form of equations 6.2 involving an iterative technique. As in any finite difference solution the accuracy depends amongst other things, on the spacing between A and B and the rate of convergence of the solution. In solving this problem numerically an initial estimate must be made of the values of x, y, σ and θ at point C. An expedient assumption is to consider that the characteristics at A and B are straight lines with slopes respectively $(\theta_A - \mu)$ and $(\theta_B + \mu)$.

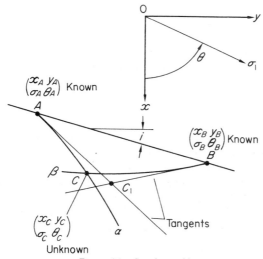

Figure 6.1   Cauchy problem

Referring to *Figure 6.1*, point C_1 is the position where tangents to the stress characteristics at A and B intersect. The values of x_{C_1} and y_{C_1} are obtained from

$$y_{C_1} - y_A = \tan(\theta_A - \mu).(x_{C_1} - x_A) \tag{6.3a}$$

$$y_{C_1} - y_B = \tan(\theta_B + \mu).(x_{C_1} - x_B) \tag{6.3b}$$

σ_{C_1} and θ_{C_1} are given by

$$(\sigma_{C_1} - \sigma_A) - 2\sigma_A.\cot 2\mu.(\theta_{C_1} - \theta_A) = a(\theta_A).(x_{C_1} - x_A) \tag{6.3c}$$

$$(\sigma_{C_1} - \sigma_B) + 2\sigma_B.\cot 2\mu.(\theta_{C_1} - \theta_B) = b(\theta_B).(x_{C_1} - x_B) \tag{6.3d}$$

Now the actual curved stress characteristics passing through A and B intersect at C. To locate C we can now use this first estimate to the values of x_C, y_C, σ_C and θ_C. To give a better estimate of the position of C, θ_{C_2} is calculated using the values of x_{C_1} and y_{C_1}. This value of θ_{C_2} is then used to calculate σ_{C_2}, x_{C_2} and y_{C_2}. The process is convergent and is thus repeated until sufficient numerical accuracy is achieved.

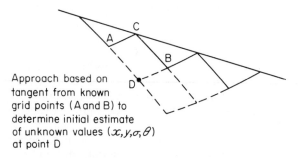

Approach based on
tangent from known
grid points (A and B) to
determine initial estimate
of unknown values (x, y, σ, θ)
at point D

Figure 6.2   Gourset problem

To extend the pattern of stress characteristics further, this procedure can be repeated as illustrated by the dotted lines in *Figure 6.2* or alternatively it can be shown that the solution is obtained more expediently by assuming that $ABCD$ in *Figure 6.2* forms a parallelogram and hence to obtain initial estimates of $x_D, y_D, \sigma_D, \theta_D$ at point D the following parallelogram relationship can be used

$$x_D = x_A + x_B - x_C$$

similarly for y_D, σ_D, and θ_D.

One special case of the Goursat problem which is of particular importance in earth pressure problems is the 'centred fan', and this field arises when there is a singularity in a stress field. Such a singularity commonly occurs in the evaluation of pressures against retaining structures. The centre 0 of such a fan will be located at the top of the wall, *Figure 6.3*.

To evaluate the stress system in a fan consider a β characteristic at an infinitesimal depth, and treat the soil above this depth as weightless. Then equation 6.2 simplifies to

$$d\sigma_p + 2\sigma_p.\cot 2\mu.d\theta = 0 \tag{6.4a}$$

In the limit the coordinates of W and S of *Figure 6.3* are identical and hence it is possible to subdivide SW into grid points with the same coordinates $(0,0)$ but with σ varying as θ varies along the β line from θ_{OB} to θ_{OA}. Integrating equation 6.3 gives,

$$\sigma_P = \sigma_{OB,p}. \exp(-2.\cot 2\mu.(\theta - \theta_{OB})) \qquad (6.4b)$$

The problem thus becomes one with known values along two characteristics and the field can be extended further taking into account the self-weight of the material.

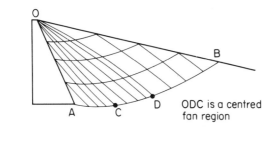

ODC is a centred fan region

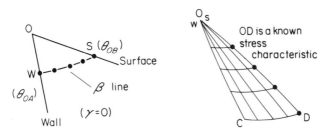

Figure 6.3   Centred fan (Gourset problem)

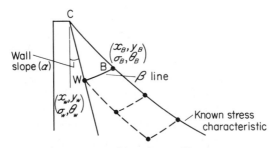

Figure 6.4   Mixed boundary problem

The mixed boundary problem of importance in the present analysis is illustrated in *Figure 6.4*. Along the wall let it be assumed that the angle, ϕ_W, between the normal and shear components is specified, thus θ_w can be calculated. From this value, and for a specific value of wall slope, α, the β characteristic commencing at point B on the known α characteristic can be established. A Goursat problem then remains to extend the field as shown by the dotted lines in *Figure 6.4*. In order to complete the stress field it is seen that it is necessary to repeat the procedure for points down the wall.

For a β line of *Figure 6.4*, the stress variation for the passive case is given by equation 6.2 and the stress characteristics are defined by equation 6.1. For the active case the relevant characteristic becomes an α line. From a knowledge of θ_w and x_B, y_B, $\sigma_{B,p}$ and θ_B at the known α characteristic, the values of $\sigma_{w,p}$, x_w, y_w on the wall can be calculated. Along the wall

$$y_w = x_w \cdot \tan \alpha$$

and taking

$$\theta_{AV} = \frac{\theta_w + \theta_B}{2}$$

then,

$$y_w - y_B = \tan(\theta_{AV} + \mu) \cdot (x_w - x_B)$$

and

$$\sigma_{w,p} - \sigma_{B,p} + (\sigma_{w,p} + \sigma_{B,p}) \cdot \cot 2\mu \cdot (\theta_w - \theta_B) = -\frac{\gamma \cdot \sin(\theta_{AV} - \mu) \cdot (x_w - x_A)}{\sin 2\mu \cdot \cos(\theta_{AV} + \mu)} \quad (6.5)$$

Solving for x_w, and $\sigma_{w,p}$ leads to the expressions

$$x_w = \frac{(y_B \cdot \cos \alpha \cdot \cos(\theta_{AV} + \mu) - x_B \cdot \cos \alpha \cdot \sin(\theta_{AV} + \mu))}{(\sin \alpha \cdot \cos(\theta_{AV} + \mu) - \cos \alpha \cdot \sin(\theta_{AV} + \mu))} \quad (6.6)$$

and

$$\sigma_{w,p} = \frac{\left(\sigma_{B,p}(1 - \cot 2\mu \cdot (\theta_w - \theta_B)) - \dfrac{\gamma \cdot \sin(\theta_{AV} - \mu) \cdot (x_w - x_B)}{\cos(\theta_{AV} + \mu)} \right)}{(1 + \cot 2\mu \cdot (\theta_w - \theta_B))} \quad (6.7)$$

6.2.2   Velocity Fields

The concept of a non-associated flow rule material is important when considering the kinematics of the problem. It will be recalled (Cox (1963), Palmer (1966), Davis (1968), Roscoe (1970)) that for a c–ϕ–γ material undergoing plastic deformation the dilatancy characteristics may be represented by the use of this model. It follows that the direction of shear failure does not coincide with the stress characteristic except in the special case where the geometric parameter ψ or v (Davis (1968), Roscoe (1970)) is equal to the strength parameter ϕ (associated flow rule material).

The correct solution for active or passive pressure can be determined from the limit theorems (Drucker (1952)) for the associated flow rule material but these theorems are not applicable to a non-associated flow rule material, and thus the uniqueness of the solution for a non-associated flow rule material cannot be established by this approach. It is important to note, however, that the unique solution for an associated flow rule material is an upper bound of the solution for a non-associated flow rule material with the same ϕ value.

The slopes of the velocity characteristics are given by equation 6.1 with ϕ replaced by ψ. This means that the velocity characteristics are inclined at an angle of $\pm(\frac{1}{4}\pi - \frac{1}{2}\psi)$ to the principal stress direction *(Figure 6.5)*. For an associated flow rule material the stress and velocity characteristics coincide, and there is a maximum deviation of the characteristics for $\psi = 0$. The latter condition requires zero incremental volume change and therefore models the critical void state.

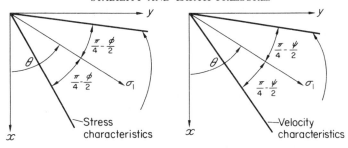

σ_1 Major principal stress
θ Major principal stress direction

Figure 6.5   Stress and velocity characteristics

The variation of velocity components along the velocity characteristics (by analogy with the stress characteristics also termed the α and β characteristics) are given by the equations

$$\frac{\partial v_x}{\partial \beta} + \tan\left(\theta + \left(\frac{\pi}{4} - \frac{\psi}{2}\right)\right) \cdot \frac{\partial v_y}{\partial \beta} = 0 \tag{6.8}$$

$$\frac{\partial v_x}{\partial \alpha} + \tan\left(\theta - \left(\frac{\pi}{4} - \frac{\psi}{2}\right)\right) \cdot \frac{\partial v_y}{\partial \alpha} = 0 \tag{6.9}$$

along β and α characteristics respectively.

The velocity characteristics can be readily derived from the stress characteristics by plotting a field with each line inclined at $\frac{1}{2}(\phi - \psi)$ to the stress line. To determine the velocity variation along the characteristics, consider the finite difference equations 6.8, 6.9.

$$\Delta v_x = -\tan\left(\theta \pm \left(\frac{\pi}{4} - \frac{\psi}{2}\right)\right) \cdot \Delta v_y \tag{6.10}$$

along a β and α line respectively.

These expressions reduce to the equations for an associated flow rule material for $\psi = \phi$.

Comparing the velocity equations with the corresponding stress equations it is seen that there are two unknowns, σ_p and θ, which vary along a stress characteristic, whereas there are three unknowns v_x, v_y, and θ along a velocity characteristic. A knowledge of the boundary values of v_x, v_y and θ, is therefore not sufficient to define the velocity characteristic. However if the velocity field is derived from the stress field, the incremental values of velocity components can be determined along a characteristic.

It was shown (Drucker (1952)) that a velocity field is kinematically admissible provided the rate of plastic work is everywhere positive. The admissibility of the velocity field can therefore be established by determining whether this criterion is satisfied at all points in the field. The rate of plastic work $\dot{W}$ in a two-dimensional field for an associated flow rule material is

$$\sigma_1 \dot{\varepsilon}_1 + \sigma_3 \dot{\varepsilon}_3 \tag{6.11}$$

where $\dot{\varepsilon}_1$ and $\dot{\varepsilon}_3$ are the principal strain increments. The rate of plastic work can be expressed in terms of velocities as,

$$\left(\frac{\partial v_x}{\partial x}+\frac{\partial v_y}{\partial y}\right)=\frac{\dot{W}}{c\cot\phi} \tag{6.12}$$

where v_x and v_y are the components of velocity in the x, y directions. The velocity components according to equations 6.8, 6.9

$$v_{xC}=-\tan\left(\frac{(\theta_A+\theta_C)}{2}-\left(\frac{\pi}{4}-\frac{\psi}{2}\right)\right)\cdot(v_{yC}-v_{yA})+v_{xA} \tag{6.13}$$

$$v_{yC}=-\tan\left(\frac{(\theta_B+\theta_C)}{2}+\left(\frac{\pi}{4}-\frac{\psi}{2}\right)\right)\cdot(v_{yC}-v_{yB})+v_{xB} \tag{6.14}$$

To establish whether the velocity field is kinematically admissible it is only necessary to investigate the sign of the left-hand side of equation 6.12 for each grid point in the velocity field.

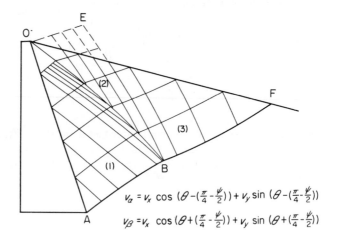

$$V_a = V_x \cos\left(\theta-\left(\frac{\pi}{4}-\frac{\psi}{2}\right)\right)+V_y \sin\left(\theta-\left(\frac{\pi}{4}-\frac{\psi}{2}\right)\right)$$

$$V_\beta = V_x \cos\left(\theta+\left(\frac{\pi}{4}-\frac{\psi}{2}\right)\right)+V_y \sin\left(\theta+\left(\frac{\pi}{4}-\frac{\psi}{2}\right)\right)$$

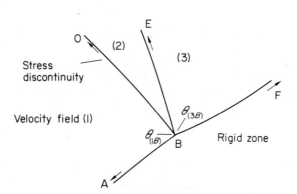

Figure 6.6   Determination of velocities in a discontinuous stress field (after Lee and Herington (1972a). Courtesy Ed. Geotechnique)

When the stress field is discontinuous† the velocity solution is similar to that for a continuous stress solution except that the stress discontinuity imposes further conditions. It can be shown that a stress discontinuity cannot be a velocity discontinuity, and also that the tangential velocity along stress discontinuity is constant and hence the rate of plastic work along a discontinuity is zero (Shield (1953)).

Consider *Figure 6.6*, point *B* is common to three separate velocity fields. If the passive case is considered as in the illustration, *AB* and *BF* are β velocity lines and v_α is continuous across *AB* and *BF*, hence $v_\alpha = 0$. Also, as *OB* is a stress discontinuity it cannot be a velocity discontinuity and v_x and v_y are continuous across it. However, θ is discontinuous across the stress discontinuity and hence v_β (*Figure 6.6*) and v_α will also be discontinuous. Values of v_x and v_y from field (1) in *Figure 6.6* can be used together with θ_{3B} from field (3) to calculate v_β and v_α in field (3) at point *B*.

Consider the line *EB*: across this line at point *B*, v_α must be discontinuous since along *BF* $v_\alpha = 0$, and hence *EB* will be a velocity discontinuity. Across this line v_β will be continuous and v_α discontinuous. Along *EB*

$$\Delta v_\alpha = v_\alpha \exp(-\cot 2\mu . (\theta - \theta_{3B})) \tag{6.15}$$

where Δv_α is equal to the change in velocity from field (2) to field (3). Equation 6.15 plus the fact that v_α is continuous across *EB*, and that the velocities along *BF* are specified, means that the velocities of field (3) can be calculated and hence the entire velocity field can be determined.

6.3 INFINITE SLOPE. NUMERICAL ANALYSIS

Equations 6.2 define the variation of mean normal programme stress, σ_p, along the stress characteristics. Let us consider the active case. For the downward slope shown in *Figure 6.1*, θ_0 is equal to $-i$ and $+i$ for an upward slope, and as an example take $i = 20°$, $\gamma = 1$, $\phi = \mu = 30°$, and unit normal stress along the boundary. Then for the β and α characteristics, respectively,

$$\Delta\sigma_p + \frac{2}{\sqrt{3}}.\sigma_p.\Delta\theta = -\frac{2}{\sqrt{3}}\frac{\sin(\theta-10)}{\cos(\theta+30)}.\Delta x \tag{6.16a}$$

$$\Delta\sigma_p - \frac{2}{\sqrt{3}}.\sigma_p.\Delta\theta = \frac{2}{\sqrt{3}}\frac{\sin(\theta+50)}{\cos(\theta-30)}.\Delta x \tag{6.16b}$$

and from equation 6.1

$$\Delta y = \Delta x . \tan(\theta \pm 30) \tag{6.17}$$

To begin the iterative cycle x_{C1}, y_{C1} of *Figure 6.7* must be calculated. Taking $(x_B, y_B) = (0,0)$, and $(x_A, y_A) = (-\sin i, \cos i)$, x_{C1}, y_{C1} can be calculated from equations 6.3a and 6.3b, to be (0.6428, 0.7660). Writing

$$R = -\frac{2}{\sqrt{3}}\frac{\sin(\theta-10)}{\cos(\theta+30)} \quad \text{and} \quad S = \frac{2}{\sqrt{3}}\frac{\sin(\theta+50)}{\cos(\theta-30)}$$

† For discussion of establishing the stress field see Lee and Herington (1972a).

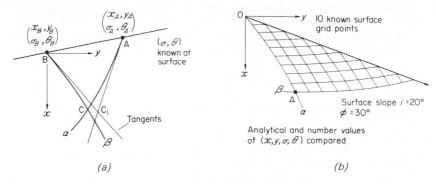

(a)   (b)

Figure 6.7   Infinite analysis
(a) Cauchy problem (b) Numerical analysis grid

Equations 6.16 become

$$\Delta\sigma_p + \frac{2}{\sqrt{3}} \cdot \sigma_p \cdot \Delta\theta = R \cdot \Delta x_{BC} \tag{6.18a}$$

$$\Delta\sigma_p - \frac{2}{\sqrt{3}} \cdot \sigma_p \cdot \Delta\theta = S \cdot \Delta x_{AC} \tag{6.18b}$$

where Δx_{BC} is the finite difference along β line, that is, $x_C - x_B$. For this example $\Delta\sigma_p$ and $\Delta\theta$ will be the same along both the α and β lines (as $\sigma_B = \sigma_A = \sigma_0$, $\theta_B = \theta_A = \theta_0$). Therefore

$$\frac{4\sigma_p}{\sqrt{3}} \Delta\theta = R\Delta x_{BC} - S\Delta x_{AC}$$

$$\Delta\theta = \frac{\sqrt{3}}{4\sigma_p}(R\Delta x_{BC} - S\Delta x_{AC}) \tag{6.19}$$

hence $$\Delta\sigma_p = R\Delta x_{BC} - \frac{2\sigma_p}{\sqrt{3}}\Delta\theta$$

Initially assuming $\theta = \theta_0 = 20°$ it is possible to calculate R and S and then $\Delta\theta = \theta_{C1} - \theta_0$ hence θ_{C1} from equation 6.19 and σ_{C1} from equation 6.20. Now having an estimate of θ_{C1} and σ_{C1} R and S can be recalculated this time using $\theta = \frac{1}{2}(\theta_{C1} + \theta_B)$. Then calculate θ_{C2} from equation 6.19 and using $\sigma_p = \frac{1}{2}(\sigma_{C1} + \sigma_B)$ in equation 6.20 calculate σ_{C2}. This new value of θ_{C2} is used to calculate x_{C2} and y_{C2} from $\Delta y = \tan(\theta \pm \mu)\Delta x$ using $\theta = \frac{1}{2}(\theta_{C2} + \theta_A$ or $\theta_B)$.

This process is then repeated until sufficient accuracy is obtained. Results of the iterative method are given in *Table 6.1*. This numerical method was continued and showed that the solution was convergent.

Programme CAUCHY was used to calculate the values of x, y, σ, θ at the intersection grid points obtained from stress characteristics from a known surface (*see Figure 6.7*).

It was also possible to substitute values of θ into the analytical expressions from σ, x, y as functions of θ for this particular case and obtain their variation along α and β characteristics. A comparison could be made at point C between the analytical values of σ, θ, x, y and the values obtained by programme CAUCHY. The comparison demonstrated the accuracy of the programme based on the finite difference approximations to the stress equations, *see Table 6.2.*

Table 6.1

Step 1.	x_{C_1} 0.642786	y_{C_1} 0.766045		
Step 2.	θ_{C_1} 0.052868	σ_{C_1} 1.538754		
Step 3.	θ_{C_2} 0.185421	σ_{C_2} 1.554505	x_{C_2} 0.676799	y_{C_2} 0.599088
Step 4.	θ_{C_3} 0.151464	σ_{C_3} 1.573203	x_{C_3} 0.667710	y_{C_3} 0.675020
Step 5.	θ_{C_4} 0.162178	σ_{C_4} 1.568647	x_{C_4} 0.670996	y_{C_4} 0.655693

Table 6.2

	θ	x	y	σ
Analytical	1.0991	2.8168	2.0079	5.5185
Programme CAUCHY	1.0991	2.8161	2.0085	5.5179

The velocity solution is not considered here although the velocities associated with these derived stress characteristics can obviously be determined by using the velocity equations. To obtain these velocities it is required to know the boundary deformation conditions.

6.4   RIGID RETAINING STRUCTURES

The types of stress fields developed in the retained mass are illustrated in *Figure 6.8* for the passive case. Rankine's stress field is correct for a horizontal fill surface and a vertical, shear-stress free wall. The value of θ at the wall, θ_w, then equals the value at the stress free surface, θ_s. The development of a shear stress acting downwards on the soil causes $\theta_w < \theta_s$ and a centred fan OAB, *Figure 6.8b* must be included in the stress field. It is considered (Lee and Herington (1972a, b)) that the limiting case occurs when the tangent to the velocity characteristic at the interface is in the interfacial direction. This is the 'rough' positive state.

Reversal of the direction of shear requires $\theta_w > \theta_s$ and a discontinuity must be introduced. *Figure 6.8c* shows the stress field and the limiting case is the 'negative' rough state. Similarly, for the active case the fan or discontinuity must be introduced according to the relative value of θ_s compared with θ_w.

Figures 6.9 and *6.10* illustrate the spectrum of stress fields for active and passive states.

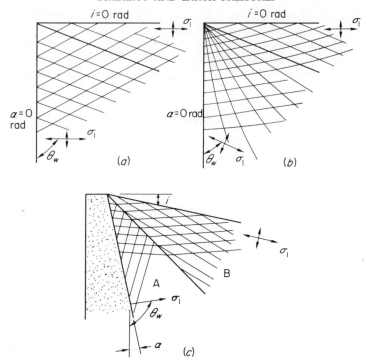

Figure 6.8   Stress fields in a retained soil mass (after Lee and Herington (1972b).Courtesy ASCE)

6.4.1   Soil-wall interface conditions

Assume that the shearing stress along the wall can be expressed by a Coulomb-type expression

$$\tau = c_w + \sigma_n . \tan \phi_w \qquad (6.21a)$$

thus, if τ_p and $\sigma_{n,p}$ are the programme shear and normal stresses, the wall boundary condition to be satisfied is,

$$\tau_p = \tau = c_w + (\sigma_{n,p} - c \cot \phi) . \tan \phi_w \qquad (6.21b)$$

The strengths τ_p and $\sigma_{n,p}$ must be expressed in terms of the variables used in the analysis, that is, σ and θ, and substituted into the boundary equation 6.21b. From resolution of forces on a small element at the wall-soil interface (Figure 6.11).

$$\sigma_{x,p} = \sigma_p(1 + \cos 2\mu . \cos 2\theta) \qquad (6.22a)$$

$$\sigma_{y,p} = \sigma_p(1 - \cos 2\mu . \cos 2\theta) \qquad (6.22b)$$

$$\tau_{xy,p} = \sigma_p . \cos 2\mu . \sin 2\theta \qquad (6.22c)$$

emphasising that the subscript p indicates that these are the stresses obtained by the plasticity solution with $c = 0$ and $\sigma_x, \sigma_y, \tau_{xy}$ are the real stresses for a

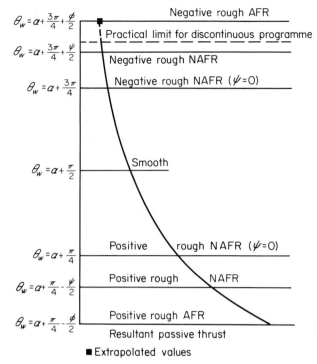

$\theta_w = \alpha + \frac{3\pi}{4} + \frac{\phi}{2}$ Negative rough AFR

Practical limit for discontinuous programme

$\theta_w = \alpha + \frac{3\pi}{4} + \frac{\psi}{2}$ Negative rough NAFR

$\theta_w = \alpha + \frac{3\pi}{4}$ Negative rough NAFR ($\psi = 0$)

$\theta_w = \alpha + \frac{\pi}{2}$ Smooth

$\theta_w = \alpha + \frac{\pi}{4}$ Positive   rough NAFR ($\psi = 0$)

$\theta_w = \alpha + \frac{\pi}{4} - \frac{\psi}{2}$ Positive rough   NAFR

$\theta_w = \alpha + \frac{\pi}{4} - \frac{\phi}{2}$ Positive rough AFR

Resultant passive thrust

■ Extrapolated values

Figure 6.9   Range of solutions for Passive Thrust Associated (AFR) and Non-Associated Flow Rule (NAFR) material (after Lee and Herington (1972b). Courtesy ASCE)

$c-\phi-\gamma$ material. Substitution into equation 6.21b gives

$$\sigma_p . \cos 2\mu . \sin 2(\theta - \alpha)$$
$$= c_w + \{\sigma_p(1 - \cos 2\mu . \cos 2(\theta - \alpha)) - c \cot \phi\} \tan \phi_w \qquad (6.23)$$

With reference to *Figure 6.11* the real boundary stresses are

$$\sigma_n = \sigma_p(1 - \sin \phi . \cos 2(\theta - \alpha)) - c \cot \phi \qquad (6.24a)$$

$$\tau = \sigma_p . \sin \phi . \sin 2(\theta - \alpha) \qquad (6.24b)$$

that is
$$\frac{\tau}{\sigma_n + c \cot \phi} = \frac{\sin \phi . \sin 2(\theta - \alpha)}{(1 - \sin \phi . \cos 2(\theta - \alpha))} \qquad (6.25a)$$

or in the form of 6.21a

$$c_w = \frac{c \cos \phi . \sin 2(\theta - \alpha)}{1 - \sin \phi . \cos 2(\theta - \alpha)} \qquad (6.26a)$$

$$\tan \phi_w = \frac{\sin \phi . \sin 2(\theta - \alpha)}{1 - \sin \phi . \cos 2(\theta - \alpha)} \qquad (6.26b)$$

where $\theta = \theta_w$, the value along the soil-wall interface.

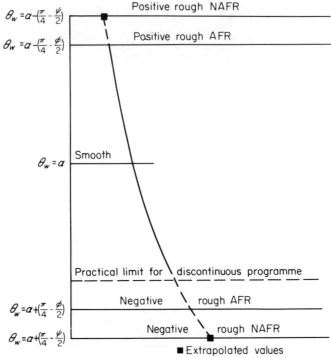

$$\theta_w = a - \left(\frac{\pi}{4} - \frac{\psi}{2}\right)$$

Positive rough NAFR

$$\theta_w = a - \left(\frac{\pi}{4} - \frac{\phi}{2}\right)$$

Positive rough AFR

$$\theta_w = a$$ Smooth

Practical limit for ⟍ discontinuous programme

Negative ⟍ rough AFR

$$\theta_w = a + \left(\frac{\pi}{4} - \frac{\phi}{2}\right)$$

Negative ⟍ rough NAFR

$$\theta_w = a + \left(\frac{\pi}{4} - \frac{\psi}{2}\right)$$

■ Extrapolated values

Figure 6.10　Range of solutions for active thrust

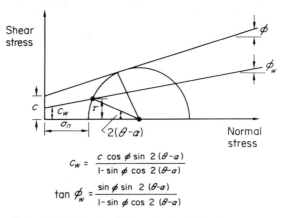

$$c_w = \frac{c \cos\phi \sin 2(\theta - a)}{1 - \sin\phi \cos 2(\theta - a)}$$

$$\tan\phi_w = \frac{\sin\phi \sin 2(\theta - a)}{1 - \sin\phi \cos 2(\theta - a)}$$

Figure 6.11　Stress conditions at soil-wall interface

For the important cases relevant to a rough interface, the values of θ are shown in *Figures 6.9, 6.10*.

$$\theta_w = \alpha + \tfrac{1}{4}\pi - \tfrac{1}{2}\psi \quad \text{positive, passive case}$$
$$\theta_w = \alpha + \tfrac{3}{4}\pi + \tfrac{1}{2}\psi \quad \text{negative, passive case}$$
$$\theta_w = \alpha - \tfrac{1}{4}\pi + \tfrac{1}{2}\psi \quad \text{positive, active case}$$
$$\theta_w = \alpha + \tfrac{1}{4}\pi - \tfrac{1}{2}\psi \quad \text{negative, active case}$$

Substitution of θ_w into equations 6.26 for the associated flow rule material ($\psi = \phi$) shows that $c_w = c$, and $\phi_w = \phi$. For the non-associated flow rule material $c_w < c$ and $\phi_w < \phi$. For the limiting case of $\psi = 0$. Then

$$c_w = |c \cos \phi|$$

$$\tan \phi_w = |\sin \phi|$$

For example, if the non-associated flow rule material is used to model a soil in the critical void state then $\psi = 0$, and taking a ϕ_{cv} of 32°, ϕ_w is seen to be 27.9°. If the peak stress state is modelled ψ for a granular soil is of the order of 10°, and taking a ϕ peak of 42°, substitution for these values into equation 6.26 shows that the corresponding value of ϕ_w is 30.6°. This provides some theoretical justification for the reduction in wall friction values universally recommended in Codes, and it is of interest to note that the commonly recommended value of $\frac{2}{3}\phi$ is close to that calculated above.

6.4.2 Active and passive thrusts

An analysis of the stress field establishes that the real normal and shear stresses along the interface can be expressed as

$$\sigma_n = A \cdot c \cdot \cot \phi + B \cdot \gamma \cdot \Delta x \qquad (6.27a)$$

$$\tau = C \cdot c \cdot \cot \phi + D \cdot \gamma \cdot \Delta x \qquad (6.27b)$$

A, B, C and D are numerical values established by the analysis. Δx is the difference in x value of a characteristic between the starting point on the free surface and the finishing point on the soil-wall interface. The total thrust on the wall can be obtained by considering a series of these characteristics, calculating the stress components from equations 6.27, and integrating numerically. The normal and tangential thrusts, N and T, are then given as

$$N = H \cdot E \cdot c \cdot \cot \phi + F \cdot \gamma \cdot H^2 \qquad (6.28a)$$

$$T = H \cdot G \cdot c \cdot \cot \phi + J \cdot \gamma \cdot H^2 \qquad (6.28b)$$

Again, the expressions contain factors (E, F, G, H) which are determined by the analysis for a particular combination of geometric parameters defining the slope of the surface and the wall.

For convenience, the wall thrusts can be expressed in dimensionless terms and related to the properties of the soil which are also expressed in a dimensionless form. For example, the total horizontal wall force per metre width of wall at passive failure, P_{PH}, can be made dimensionless by dividing by γH^2 or by $cH \cot \phi$, and this can be plotted against $c \cot \phi / \gamma H$ or $\gamma H / c \cot \phi$ respectively. H is the vertical component of wall height. Thus,

$$\frac{P_{PH}}{\gamma H^2} = E \cdot \frac{c \cot \phi}{\gamma H} + F \qquad (6.27a)$$

or,

$$\frac{P_{PH}}{Hc \cot \phi} = E + F \cdot \frac{\gamma H}{c \cot \phi} \qquad (6.27b)$$

This approach avoids the necessity for evaluating the failure states for all values of γ and $c \cot \phi$, and these can be conveniently taken as unity. The generalised solution for a specific wall and surface geometry is represented in *Figure 6.12*.

When $\gamma = 0$,

$$P_{PH} = E . H . c . \cot \phi = H(N_{QPH}) \qquad (6.28a)$$

and when $c = 0$,

$$P_{PH} = F . \gamma H^2 = H\left(\frac{\gamma H}{2} . N_{\gamma PH}\right) \qquad (6.28b)$$

For intermediate values of c and γ, the generalised expression for passive thrust is

$$P_{PH} = fH\left(N_{QPH} . c + \frac{\gamma H}{2} . N_{\gamma PH}\right) \qquad (6.29)$$

where f is a factor equal to unity for the conditions expressed by equations 6.28. This expression is applicable to both the horizontal and vertical force components and to both the active and passive states by rewriting equation with the appropriate subscripts (P passive, A active, H horizontal, V vertical).

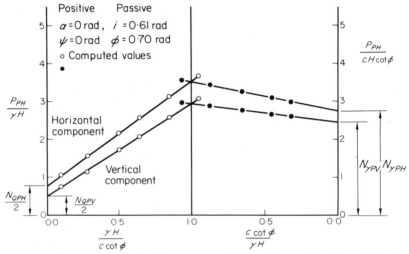

Figure 6.12   Typical plot for the determination of thrust factors (after Lee and Herington (1972b). Courtesy ASCE)

The factor f is a function of all of the variables involved but it is usually very close to unity. It increases with ϕ and with wall friction but decreases with wall angle α and slope i. For a vertical wall, horizontal surface, and a ϕ of 0.70 rad, f ranges from 0.995 to 1.036 for values of ϕ_w from 0° to 0.70 rad—taking into account the value of ϕ_w likely to be developed, the value would be closely equal to 1.02. The maximum deviation from unity for i values of ± 0.35 rad, vertical wall ($\phi = 0.79$ rad), is 1.09.

Thus for practical purposes the simplified superposition form can be used, viz.,

$$P = H\left(N_Q . c + \frac{\gamma H}{2} . N_\gamma\right) \qquad (6.30)$$

with the appropriate subscripts.

By considering the statics of an element along the soil-wall interface it is possible to relate the vertical and horizontal thrust factors for the fully rough interfacial state. For the positive passive state, *(Figure 6.12)*.

$$N_{QPV} = -N_{QPH} \tan \alpha - \frac{c_w}{c}(1+\tan^2 \alpha) \qquad (6.31a)$$

$$N_{\gamma PV} = N_{\gamma PH}\left(\frac{\tan \phi_w \cos \alpha - \sin \alpha}{\tan \phi_w \sin \alpha + \cos \alpha}\right) \qquad (6.31b)$$

The vertical force component acting on the wall is deemed positive if it acts in the upward direction. The expressions also apply to the negative active state, if the P subscript is replaced by A, and both the negative passive and the positive active states are obtained by changing the sign of c_w and $\tan \phi_w$.

Figure 6.13a shows N_{QAH} as a function of ϕ (0.35–0.79 rad) for surface slopes of ±0.35 rad, and wall angle α (0.035 rad). In this plot, and in the corresponding plot for $N_{\gamma AH}$ *(Figure 6.13b)* the differences between the factors for an associated and non-associated flow rule material are insignificant. The interfacial state is fully rough, positive.

Figure 6.14 shows the corresponding results for the passive state for a surface slope of +0.35 rad. The passive values are detailed in *Tables 6.3* ($i = 0$ rad), 6.4 ($i = 0.35$ rad), 6.5 ($i = +0.61$ rad) and 6.6 ($i = -0.35$ rad). Both the associated flow rule values ($\phi_w = \phi$) and the particular non-associated flow rule value $\psi = 0$ ($\tan \phi_w = \sin \phi$) are given.

The factors N_{QPM}, $N_{\gamma PM}$ define the moment of the passive thrust about the toe of the wall. The relevant equation analogous to equation 6.30 is

$$M_P = H^2\left(c \cdot N_{QPM} + \frac{\gamma H}{3} \cdot N_{\gamma PM}\right)$$

The effect of ψ is reflected in the value of ϕ_w. *Figure 6.15* shows the effect of the wall friction in more detail. Some independent checks on the values are noted in this figure.

In general the Coulomb analysis for active pressure leads to a solution close to the lower bound plasticity value for a rough interface provided the value of wall friction in the Coulomb analysis is taken as (approximately) $\frac{2}{3}\phi$. This conclusion is correct for both continuous and discontinuous stress fields. Studies (Lee and Herington (1972a)) have also indicated that the passive pressure calculations based on a logarithmic failure surface are likely to be close (within 10%) to the plasticity solution (associated flow rule material) for the *same* angle of wall friction.

If the logarithmic spiral analysis is carried out for a wall friction of $\frac{2}{3}\phi$, the resultant thrust is considerably less than the value obtained by the plasticity theory for an associated flow rule material and a rough soil-wall interface. For example, consider the case of $\alpha = 0$ rad, $i = 0$ rad, $\phi = 0.70$ rad, $c = 0$, then the resultant thrust for the passive state is,

$$P_P = \tfrac{1}{2}\gamma H^2 N_{\gamma P} \qquad (6.32)$$

where $N_{\gamma P}$ is the vector sum of $N_{\gamma PV}$ and $N_{\gamma PH}$. For a rough interface the plasticity values for $N_{\gamma P}$ are 18.5 for $\psi = \phi$ and 12.2 for $\psi = 0$. The value obtained by the logarithmic spiral method is 13.5 and thus happens to be quite close to the correct value for a realistic ψ.

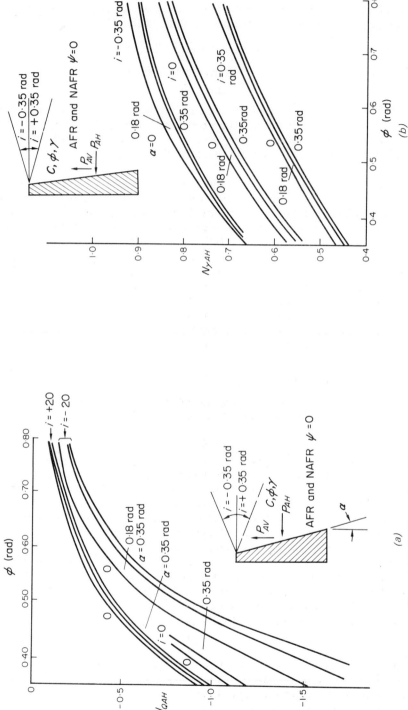

Figure 6.13(a) Thrust factor N_{QAH}. Active state. Rough. Positive (after Lee and Herington (1972b). Courtesy ASCE)
(b) Thrust factor $N_{\gamma AH}$. Active state. Rough. Positive (after Lee and Herington (1972b). Courtesy ASCE)

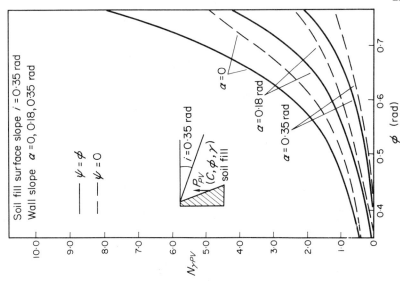

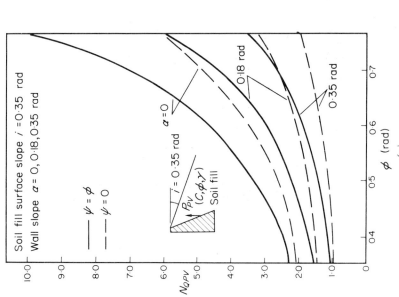

Figure 6.14(a) Passive thrust factor N_{QPV}. Rough interface, $i = 0.35$ rad, $\Psi = 0$, ϕ. Positive (after Lee and Herington (1972a). Courtesy Ed. Geotechnique)
(b) Passive thrust factor $N_{\gamma PV}$. Rough interface, $i = 0.35$ rad, $\Psi = 0$, ϕ. Positive

Table 6.3 THRUST FACTORS. ROUGH INTERFACE. POSITIVE PASSIVE STATE.
$i = 0$ rad (after Lee and Herington (1972a), courtesy Ed. Geotechnique).

Wall slope α (rad)	Friction angle (rad)	Type of field		Passive thrust coefficients											
				N_{QPV}		N_{QPH}		N_{QPM}		$N_{\gamma PV}$		$N_{\gamma PH}$		$N_{\gamma PM}$	
		$\psi=\varphi$	$\psi=0$	$\psi=\varphi$	$\psi=0$	$\psi=\varphi$	$\psi=0$	$\psi=\varphi$	$\psi=0$	$\psi=\varphi$	$\psi=0$	$\psi=\varphi$	$\psi=0$	$\psi=\varphi$	$\psi=0$
0 (0°)	0.39	C	C	3.15	2.90	5.10	4.90	2.50	2.40	1.40	1.35	3.45	2.35	1.80	1.75
	0.52	C	C	5.00	4.20	6.95	6.40	3.45	3.15	3.30	2.60	5.65	5.00	3.00	2.70
	0.70	C	C	11.1	7.10	12.2	9.65	6.10	4.80	12.0	6.60	14.2	10.2	7.20	5.10
	0.79	C	C	18.0	11.9	17.0	15.3	8.55	7.60	26.0	14.9	26.0	21.2	14.1	11.0
0.18 (10°)	0.39	C	C	2.05	1.80	4.65	4.45	2.05	1.95	0.65	0.60	3.00	2.95	1.40	1.40
	0.52	C	C	3.20	2.75	6.05	5.70	2.65	2.45	1.80	1.50	4.95	4.75	2.35	2.30
	0.70	C	C	6.85	4.50	1.01	9.30	4.30	3.75	6.55	4.50	11.0	9.10	5.00	4.60
	0.79	C	C	10.9	6.30	14.0	12.1	5.75	5.20	13.2	7.50	18.8	15.2	8.25	7.05
0.35 (20°)	0.39	C	C	1.15	1.00	4.10	4.05	1.80	1.75	0.15	0.05	2.75	2.65	1.35	1.35
	0.52	C	C	1.85	1.45	5.30	5.00	2.25	2.20	0.75	0.45	4.25	4.00	2.05	1.90
	0.70	C	C	4.05	2.50	8.60	7.40	3.35	3.10	3.20	2.40	8.80	7.40	4.00	3.70
	0.79	C	C	6.00	3.20	11.3	9.30	4.30	4.00	6.55	3.40	14.2	11.3	6.00	5.10

Note: C = continuous field.

Table 6.4 THRUST FACTORS. ROUGH INTERFACE. POSITIVE PASSIVE STATE.
$i = 0.35$ rad (after Lee and Herington (1972a), courtesy Ed. Geotechnique).

Wall slope α (rad)	Friction angle (rad)	Type of field		Passive thrust coefficients											
				N_{QPV}		N_{QPH}		N_{QPM}		$N_{\gamma PV}$		$N_{\gamma PH}$		$N_{\gamma PM}$	
		$\psi=\varphi$	$\psi=0$	$\psi=\varphi$	$\psi=0$	$\psi=\varphi$	$\psi=0$	$\psi=\varphi$	$\psi=0$	$\psi=\varphi$	$\psi=0$	$\psi=\varphi$	$\psi=0$	$\psi=\varphi$	$\psi=0$
0 (0°)	0.39	C	C	2.35	1.90	3.20	3.10	1.60	1.55	0.60	0.50	1.35	1.30	0.65	0.60
	0.52	C	C	3.40	2.95	4.10	3.85	2.05	1.90	1.30	1.00	2.25	2.15	1.15	1.10
	0.70	C	C	6.15	4.50	6.10	5.75	3.05	2.85	3.95	2.65	4.70	4.15	2.40	2.10
	0.79	C	C	9.00	5.65	8.00	7.25	4.00	3.60	7.50	4.70	7.50	6.40	3.75	3.20
0.18 (10°)	0.39	C	C	1.60	1.30	2.85	2.75	1.25	1.25	0.25	0.20	1.15	1.10	0.55	0.55
	0.52	C	C	2.30	1.75	3.50	3.40	1.55	1.45	0.70	0.55	1.90	1.70	0.85	0.80
	0.70	D	C	3.95	2.60	5.10	4.55	2.10	2.00	2.10	1.35	3.60	3.25	1.75	1.50
	0.79	D	C	5.50	3.55	6.40	5.75	2.70	2.40	3.80	2.40	5.40	4.50	2.40	2.15
0.35 (20°)	0.39	D	C	1.05	0.90	2.45	2.40	0.95	0.95	0.05	0.00	1.10	1.05	0.50	0.40
	0.52	D	C	1.45	1.25	3.00	2.95	1.20	1.15	0.30	0.25	1.55	1.65	0.70	0.65
	0.70	D	C	2.35	1.50	4.15	3.60	1.60	1.50	1.00	0.60	2.80	2.45	1.25	1.10
	0.79	D	C	3.20	1.85	5.10	4.10	1.90	1.70	1.85	1.15	4.00	3.50	1.75	1.50

Note: D = discontinuous field.

Table 6.5 THRUST FACTORS. ROUGH INTERFACE. POSITIVE PASSIVE STATE. $i = 0.61$ rad (after Lee and Herington (1972a), courtesy Ed. Geotechnique).

Wall slope α (rad)	Friction angle (rad)	Type of field		Passive thrust coefficients											
				N_{QPV}		N_{QPH}		N_{QPM}		$N_{\gamma PV}$		$N_{\gamma PH}$		$N_{\gamma PM}$	
		$\psi=\varphi$	$\psi=0$	$\psi=\varphi$	$\psi=0$	$\psi=\varphi$	$\psi=0$	$\psi=\varphi$	$\psi=0$	$\psi=\varphi$	$\psi=0$	$\psi=\varphi$	$\psi=0$	$\psi=\varphi$	$\psi=0$
0 (0°)	0.66	D	C	3.55	2.65	3.30	3.10	1.60	1.45	1.05	0.85	1.35	1.35	0.70	0.65
	0.70	D	C	4.00	2.90	3.50	3.25	1.80	1.60	1.30	0.95	1.50	1.45	0.80	0.75
	0.79	D	C	5.30	3.40	4.30	3.80	2.10	1.90	2.35	1.45	2.35	2.20	1.20	1.10
0.18 (10°)	0.66	D	C	2.35	1.70	2.65	2.45	1.10	1.05	0.55	0.35	1.05	0.85	0.50	0.40
	0.70	D	C	2.55	1.75	2.85	2.50	1.20	1.10	0.70	0.50	1.15	1.10	0.60	0.45
	0.79	D	C	3.25	2.00	3.45	2.90	1.35	1.20	1.25	0.65	1.75	1.45	0.80	0.60
0.35 (20°)	0.66	D	D	1.55		2.10		0.75		0.25		0.75		0.35	
	0.70	D	D	1.70		2.20		0.80		0.30		0.85		0.35	
	0.79	D	D	2.10		2.60		0.90		0.60		1.20		0.45	

Table 6.6 THRUST FACTORS. ROUGH INTERFACE. POSITIVE PASSIVE STATE.
$i = -0.35$ rad (after Lee and Herington (1972a), courtesy Ed. Geotechnique).

Wall slope α (rad)	Friction angle (rad)	Type of field		Passive thrust coefficients											
				N_{QPV}		N_{QPH}		N_{QPM}		$N_{\gamma PV}$		$N_{\gamma PH}$		$N_{\gamma PM}$	
		$\psi=\varphi$	$\psi=0$	$\psi=\varphi$	$\psi=0$	$\psi=\varphi$	$\psi=0$	$\psi=\varphi$	$\psi=0$	$\psi=\varphi$	$\psi=0$	$\psi=\varphi$	$\psi=0$	$\psi=\varphi$	$\psi=0$
0 (0°)	0.39	C	C	4.15	3.85	7.85	7.75	3.65	3.60	2.30	2.15	5.60	5.60	2.80	2.75
	0.52	C	C	7.65	6.60	11.45	11.30	5.55	5.40	6.45	5.50	11.20	10.70	5.60	5.40
	0.70	C	C	20.1	14.7	22.8	21.2	11.3	10.2	29.6	19.8	35.0	31.2	17.5	15.6
0.18 (10°)	0.39	C	C	2.50	2.20	7.00	6.90	3.15	3.05	1.20	1.00	5.20	5.05	2.50	2.45
	0.52	C	C	4.05	3.80	10.05	9.90	4.75	4.30	3.60	2.80	9.60	9.40	5.40	4.40
	0.70	C	C	12.2	8.10	19.5	17.4	8.35	7.85	15.8	10.1	27.4	23.6	12.9	11.2
0.35 (20°)	0.39	C	C	1.25	1.00	6.30	6.15	2.70	2.70	0.25	0.05	4.80	4.70	2.40	2.30
	0.52	C	C	2.60	1.75	8.85	8.50	3.80	3.65	1.50	0.90	8.50	8.00	4.20	3.80
	0.70	C	C	6.80	4.00	16.1	14.3	6.55	6.25	8.00	4.20	21.7	18.7	9.55	8.50

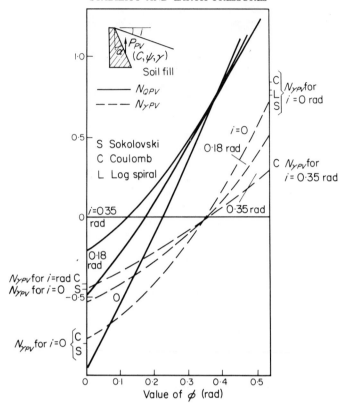

Figure 6.15   Effect of wall friction on passive thrust factors. α = 0.35 rad, φ = 0.52 rad. Positive (after Lee and Herington (1972a). Courtesy Ed. Geotechnique)

The evidence to date suggests that the traditional approximate methods based on an $\phi_w = \frac{2}{3}\phi$ yield values close to the correct lower bound solutions for realistic dilatancy rates.

Rigid wall movements leading to the active or passive states can be considered to be translation in a direction z relative to the horizontal, rotation about the top of the wall, and rotation about the base of the wall or a combination of rotation and translation. A consideration of the sign of $\dot{W}$ (equation 6.12) for a specific type of wall movement enables the uniqueness of the solution for an associated flow rule material to be evaluated for that particular wall movement.

For the movements leading to a positive passive state it was established (Lee and Herington (1972a)) that the velocity field derived from the continuous stress field was kinematically admissible provided the movement normal to the wall at the top was greater than or equal to the normal movement at the base. This did not apply to rotation about the top. The solutions for discontinuous stress fields were kinematically admissible provided the normal movement at the top of the wall exceeded the normal base movement. There is some doubt whether the solution is admissible for translation.

It was established (Lee and Herington (1972b)) that the velocity fields derived from the stress fields for the positive active case are kinematically acceptable for translation and rotation about the base, and are therefore correct solutions for an associated flow rule material.

6.4.3   Influence of wall movement

It is now relevant to discuss the relationship between the direction of wall translation and the active and passive thrusts. Consider a vertical wall moving into the soil at an angle z relative to the horizontal. The range of passive interfacial states is obtained by a rotation of the velocity characteristics.

From *Figure 6.16a*, and assuming θ_w is a constant along the wall-soil interface

$$\theta_w = \tfrac{1}{4}\pi - \tfrac{1}{2}\psi - z$$

A shear stress-free state requires $\theta_w = \tfrac{1}{2}\pi$ and thus $z = -(\tfrac{1}{4}\pi + \tfrac{1}{2}\psi)$ for z positive clockwise relative to the horizontal. This means that it is necessary for the wall to move in a direction $(\tfrac{1}{4}\pi + \tfrac{1}{2}\psi)$ upwards and towards the fill in order to develop the stress-free state along the interface. Similarly the corresponding active pressure state is developed when the wall is moved

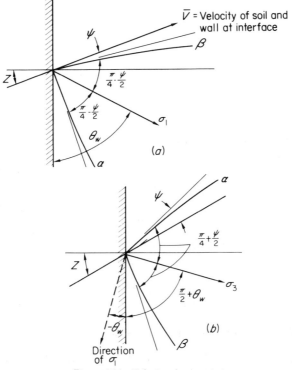

Figure 6.16   *Velocity characteristics*
(a) Passive   (b) Active

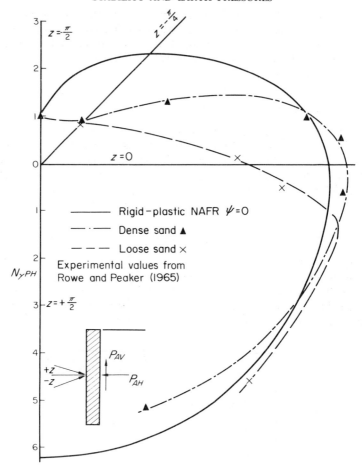

Figure 6.17 Polar representation of passive thrust at large wall movements. Vertical wall.
Direction of movement = z (after Lee and Herington (1972(b)). Courtesy ASCE)

downwards and away from the soil in a direction of $(\frac{1}{4}\pi + \frac{1}{2}\psi)$ to the horizontal.

Figure 6.17 shows a polar plot of the passive thrust for $\psi = 0$ and it is seen that the positive passive state is developed by a downward translation of the wall $(0 \leqslant z \leqslant \frac{1}{2}\pi)$. The thrust diminishes as z becomes negative and is a minimum for the negative rough passive state $(z = \frac{1}{2}\pi)$. *Figure 6.19* shows the corresponding plot for the active state with $\psi = 0$. When the material dilates $(\psi > 0)$ it is theoretically possible for a passive state to be developed for z slightly exceeding $\mp\frac{1}{2}\pi$.

Figure 6.18 shows the theoretically derived value of ϕ_w for $-\frac{1}{2}\pi \leqslant z \leqslant +\frac{1}{2}\pi$ and $\psi = 0$ (passive) and the active plot is shown as *Figure 6.20*.

6.4.4   Applicability of rigid-plastic model

Recent elegant experimental studies (Rowe and Peaker (1965), Rowe (1969), James and Bransby (1970), Narain *et al.* (1969)) has provided some data on the applicability of the rigid plastic solutions when applied to a real soil.

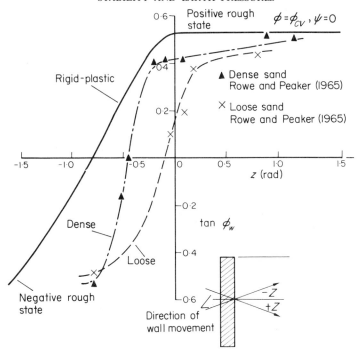

*Figure 6.18 Development of wall friction as a function of direction of movement. Passive state
(after Lee and Herington (1972b). Courtesy ASCE)*

There is now a great deal of evidence to suggest that the stress characteristics and the rupture surfaces do not coincide. From displacement
measurements in a passive field generated by rotating a rigid plate about the
base James and Bransby (1970) were able to plot lines of zero extension.
These lines can be recognised as the velocity characteristics of a non-
associated flow rule material, and coincide with the rupture surfaces. The
observed angle of intersection of the rupture surfaces to the surface was
found for dense sand to be 0.61 rad ($\phi = 0.85$ rad, $\psi = 0.35$ rad) and this is
equal to ($\frac{1}{4}\pi$ rad $-\frac{1}{2}\psi$) as predicted. For the sand in the loose state $\psi \simeq 0$ and
for the ϕ_{cv} value (0.61 rad) the angle of intersection of the rupture surfaces
should be close to $\frac{1}{4}\pi$ rad. This was observed.

Use of equation 6.26b leads to a predicted ϕ_w of 0.84 rad for the sand in
the dense state, and 0.52 rad for the loose sand. These values are in agreement
with measured values. Recalling that the rough interface concept requires
the interface to be a tangent to one set of velocity characteristics the major
principal stress direction should be ($\frac{1}{4}\pi - \frac{1}{2}\psi$) to the interfacial direction. The
observed major principal strain directions for the two density states were
closely equal to this angle ($\frac{1}{4}\pi$ rad and 0.61 rad respectively) and therefore
supports the concept that the principal axes of stress and strain coincide.

There is thus some strong quantitative evidence of the non-coincidence
of the stress characteristics and the rupture surfaces (velocity characteristics)
as predicted by plasticity theory for a rigid plastic non-associated flow rule
material.

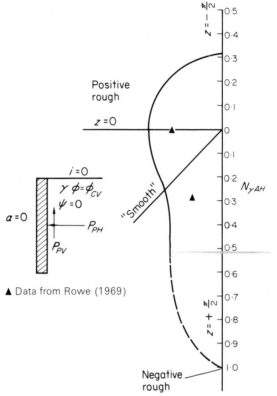

Figure 6.19   Polar representation of active thrust at large wall movements. Vertical wall

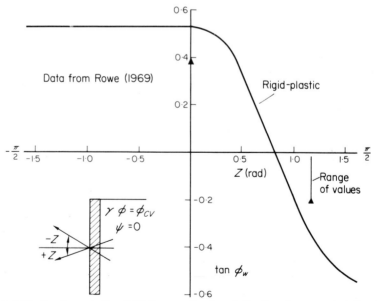

Figure 6.20   Development of wall friction as a function of direction of movement. Active state

Now let us consider the concordance of predicted and measured pressures and thrusts on a vertical rigid wall moving into a sand mass. Rowe and Peaker (1965) (see preceding volume, Chapter 7) measured the thrust components in the direction of the interface and at right angles to the wall. The soil surface was horizontal and the direction of wall movement was z to the horizontal.

The pressure exerted on the wall was proportional to depth as predicted by the plasticity solutions. This was also confirmed by Narain et al. (1969) for both translation and rotation about the base. At small rotations a parabolic distribution is observed (Rowe (1952)) but the hydrostatic distribution gradually develops as the rotation is increased. (Narain et al. (1969).)

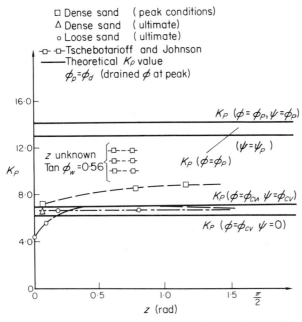

Figure 6.21 *Passive pressure coefficient K_P ($= N_{\gamma PH}$). Concordance of experimental and theoretical values (after Lee and Herington (1972a). Courtesy Ed. Geotechnique)*

The theoretical and experimental data for the passive state is shown in *Figure 6.21*. For the sand in the dense state, the values of peak ϕ_d and the corresponding ψ are 0.73 rad and 0.18 rad, respectively. At the ultimate state ϕ_{cv} was 0.56 rad–0.57 rad and $\psi = 0$. From the data previously tabulated (*Table 6.3*) the values of K_{PH} ($N_{\gamma PH}$ of equation 6.32) are 12.5 and 6.2, respectively. The peak experimental value is seen in *Figure 6.21* to increase to approximately 9 as the wall movement angle z is increased. At the stage when the recorded thrust is a maximum the peak value of ϕ_d is not simultaneously developed at all points along the surfaces thus the theoretical value of K_{PH} cannot be achieved. As the displacement of the wall is increased the ϕ values along the surfaces would approach ϕ_{cv} and the plasticity solution for $\psi = 0$ is seen to agree quite closely with measured values irrespective of the initial density state.

Figure 6.17 extends the representation to z within the range $\pm\frac{1}{2}\pi$ and corresponds to large displacements of the wall and thus the relevant value of ψ is zero. At large values of $+z$ the horizontal component of thrust was approximately constant for both the initially loose and the initially dense sand. For the initially loose sand the K_P value rapidly diminished as z was decreased below 0.18 rad whereas the K_P value for the initially dense state remained relatively constant until z became equal to -0.09 rad.

It will be shown below that the shear stress-free interfacial state requires a vertically upward component of wall movement. The value of z for a $\psi = 0$ rigid-plastic material is $-\frac{1}{4}\pi$ rad whereas the experimental values of z to reach this state were found to be $-3°$ (initially loose) and $-25°$ (initially dense). The reason for the different values is the finite amount of strain necessary to develop the shearing stress along the soil-wall interface.

As $-z$ was further increased the experimental value of K_{PH} appeared to asymptote to the theoretical value of 1.1. This state was reached at z values of -0.35 rad (initially loose) and $-\frac{1}{4}\pi$ rad (initially dense). The corresponding value for the rigid-plastic material was $-\frac{1}{2}\pi$ rad.

Adopting the kinematic boundary condition along the interface the vertical component of thrust (wall friction) can be predicted. It will be recalled that the traditional analyses require ϕ_w to be assumed. *Figures 6.18* and *6.21* show the values of ϕ_w for the rigid-plastic material with $\psi = 0$ and the corresponding experimental values—as published by Rowe and Peaker (1965) and Rowe (1969). This data clearly shows that the rigid-plastic, $\psi = 0$ model is applicable provided the ultimate state is reached along the failure surfaces and between the soil and the wall.

The solutions obtained for translation of the wall, and for rotation about the base, are the unique solutions for an idealised associated flow rule material, and are upper bound solutions for a non-associated flow rule material. The theoretical solution for the idealised material with $\psi = 0$ will be equal to or less than the value for $\psi = \phi_{cv}$, thus if the soil in the ultimate state can be represented as an idealised $\psi = 0$ material the experimental values should be below the theoretical values for $\phi = \phi_{cv}, \psi = \phi_{cv}$, not above the theoretical value as found by Narain *et al.* (1969). It can not be assumed, however, that the plastic strains at the apparently ultimate state have been sufficient to develop a critical void state in all regions. The measured thrust would exceed the theoretical value if local strains were insufficient to decrease the ϕ at all points in the plastic field to ϕ_{cv} and it is concluded that the excessive K_p values already noted for the investigations of Narain were due to this effect. The experiments of Rowe and Peaker were continued until the K_p values for initially dense and initially loose sands were virtually identical, suggesting a completely critical void state had been achieved.

K_p values obtained by Narain and his colleagues for rotation about the base were slightly higher than the values associated with wall translation. The limiting values recorded were $K_p = 9.5$ for dense sand and $K_p = 8.8$ for loose sand. The difficulty here, as in the translatory case, is the exact interpretation of the published values. In both cases the K_p values were rapidly decreasing with movement and, presumably, would have further decreased and reached the ultimate state if the experiments had been continued.

Data on the active state are more limited but generally confirm the evidence discussed above for the passive state. *Figures 6.19* and *6.20* show the active pressure results. The experimental data is quoted by Rowe (1969).

It seems clear that plasticity solutions for a rigid-plastic non-associated flow rule material with $\psi = 0$ are applicable to a real soil provided the critical void state has been developed. The direction of rupture surfaces appears to be correctly predicted at earlier stages as well as at ultimate but the wall thrust is less than the predicted peak value because of progressive rupture.

The plasticity solutions apply to wall translation and to rotation about the base.

Further understanding of the stresses and displacements in the soil mass requires a consideration of the progressive development of the failure states.

This is briefly considered in the following section.

6.4.5 Non-linear analyses

Reference has been made (Chapter 3, Section 3.4.5) to the investigations aimed at establishing a suitable stress-strain soil model. Analyses based on these or other non-linear constitutive equations have been carried out to predict the load-settlement relationships of a foundation (for example Girijavallabhan (1968), D'Appolonia and Lambe (1970)) with encouraging results. Similarly the thrust-displacement relationship for a retaining structure has been studied (Girijavallabhan (1968), Clough and Duncan (1971)) and one of the most significant advances resulting from this approach is the opportunity to investigate the progressive development of earth pressures during construction. If the analyses are adequate the stress field should approach that predicted by the plasticity theory.

Clough and Duncan (1971) used the non-linear relationship derived earlier (Duncan and Chang (1970)) to predict the thrust on a rigid retaining structure for wall translation and rotation. The particular expression for tangent modulus under a monotonic loading condition was

$$E_t = \left(1 - \frac{R_f(1-\sin\phi).(\sigma_1-\sigma_3)}{2c\cos\phi + 2\sigma_3\sin\phi}\right) K p_a \left(\frac{\sigma_3}{p_a}\right)^n \tag{6.33}$$

where R_f is the ratio of ultimate to shear stress at failure;

c, ϕ are the Coulomb parameters;

K is the modulus number;

n is the modulus exponent;

p_a is atmospheric pressure.

An outstanding feature of the analysis is the prediction of the thrust-displacement relationship leading to the active and passive states. It has been long established that the displacement or rotation causing an active state is only a fraction of the value to cause a passive state. This feature is correctly predicted.

BIBLIOGRAPHY

CLOUGH, G. W., and DUNCAN, J. M. (1971). 'Finite Element Analyses of Retaining Wall Behaviour', *Jour. Soil Mech. and Fdn. Eng. Div., ASCE,* **97,** SM 12, 1657–1673.

COX, A. D. (1963). 'The Use of Non-Associated Flow Rules in Soil Plasticity', *RARDE Rep.* (B) 2/63.

DAVIS, E. H. (1968). *Soil Mechanics Selected Topics* (Ed. I. K. Lee), Chapter 6, Butterworths, London.

D'APPOLONIA, D. J., and LAMBE, T. W. (1970). 'Method for Predicting Initial Settlement', *Jour. Soil Mech. and Fdn. Div., ASCE,* **96,** SM 2, 523–544.

DRUCKER, D. C., PRAGER, W., and GREENBERG, H. J. (1952). 'Extended Limit Design Theorems for Continuous Media', *Jour. App. Math.,* **9,** 381–389.

DRUCKER, D. C., and PRAGER, W. (1952). 'Soil Mechanics and Plastic Analyses or Limit Design', *Quart. Jour. Appl. Maths.,* **10,** 157–165.

GIRIJAVALLABHAN, C. V. (1968). 'Finite-Element Method for Problems in Soil Mechanics', *Jour. Soil Mech. and Fdn. Eng. Div., ASCE,* **94,** SM 2, 473–495.

HERINGTON, J. R. (1969). 'Application of Earth and Rock Fill for Strengthening Existing Gravity Dams', *M. Eng. Sc. Thesis,* University of Sydney, Australia.

JAMES, R. G., and BRANSBY, P. L. (1970). 'Experimental and Theoretical Investigations of a Passive Earth Problem', *Geotechnique,* **20,** 1, 17–37.

JAMES, R. G., and BRANSBY, P. L. (1971). 'A Velocity Field for Some Passive Earth Pressure Problems', *Geotechnique,* **21,** 61–83.

LEE, I. K., and HERINGTON, J. R. (1972a). 'A Theoretical Study of the Pressures Acting on a Rigid Wall by a Sloping Earth or Rock Fill', *Geotechnique,* **22,** 1, 1–26.

LEE, I. K., and HERINGTON, J. R. (1972b). 'Effect of Wall Movement on Active and Passive Pressures', *Jour. Soil Mech. and Fdn. Eng. Div., ASCE,* **98,** SM 6, 625–639.

NARAIN, J., SWAMI, S., and NANDAKUMARON, P. (1969). 'Model Study of Passive Pressure in Sand', *Jour. Soil Mech. and Fdn. Eng. Div., ASCE,* **95,** SM 4, 969–983.

PALMER, A. C. (1966). 'A Limit Theorem for Materials with Non-associated Flow Laws', *Journ. de Mècanique,* **5,** 2, 217–222.

PRAGER, W., and HODGE, P. G. (1951). *Theory of Perfectly Plastic Solids,* J. Wiley & Sons, New York.

ROSCOE, K. H. (1970). Tenth Rankine Lecture: 'The Influence of Strains in Soil Mechanics', *Geotechnique,* **20,** 2, 129–170.

ROWE, P. W. (1969). 'Progressive Failure and Strength of Sand Mass', *Proc. 7th Int. Conf. Soil Mech. and Fdn. Eng.,* **1,** 341–349.

ROWE, P. W. (1952). 'Anchored Sheet-Pile Walls', *Proc. I.C.E.,* Part 1, **1,** 27–70.

ROWE, P. W., and PEAKER, K. R. (1965). 'Passive Earth Pressure Measurements', *Geotechnique,* **15,** 1, 57–79.

SHIELD, R. T. (1953). 'Mixed Boundary Value Problems in Soil Mechanics', *Quart. Jour. Appl. Maths.,* **11,** 1, 61–74.

SOKOLOVSKI, V. V. (1956). *Statics of Soil Media,* Butterworths, London.

Chapter 7

Some Recent Developments in the Theoretical Analysis of Pile Behaviour

H. G. Poulos

7.1   INTRODUCTION

Over the last decade, the analysis and design of pile foundations behaviour has advanced from being essentially an art, supported by relatively crude computations, towards being a science, employing sophisticated analytical techniques and requiring the use of computers. Two of the aspects of pile behaviour which have been studied in considerable detail are the settlement of single piles and pile groups, and the dynamics of pile driving using the wave equation. While much has been published on the theoretical details of these aspects, there still remains an understandable reluctance on the part of the practising foundation engineer to accept these theories because of the theoretical complications involved, the apparent necessity of a computer for all calculations and the general uncertainty regarding the applicability of theory to practical piling problems.

In this chapter, an attempt is made to briefly summarise some of the more recent developments in pile settlement theory and the wave equation analysis of pile driving, to present some parametric solutions in readily-usable graphical form and to illustrate some cases in which the theoretical solutions have been applied to practical problems. Emphasis will be placed on results which have relevance for the practical foundation engineer.

7.2   PILE SETTLEMENT THEORY

7.2.1   Theoretical Approaches

The theoretical approaches for settlement analysis of piles may be classified into three broad categories:

(1) Methods based on the theory of elasticity and which employ the

equations of Mindlin (1936) for subsurface loading within a semi-infinite mass.

(2) Step-integration methods, which use measured relationships between pile resistance and pile movement at various locations along the pile.

(3) Numerical methods, and in particular, the finite element method.

Methods using the Mindlin equations.   This approach has been employed by several investigators, e.g. D'Appolonia and Romualdi (1963), Thurman and D'Appolonia (1965), Salas and Belzunce (1965), Poulos and Davis (1968), Mattes and Poulos (1969), Poulos and Mattes (1969). In all these approaches, the pile is divided into a number of uniformly-loaded elements, and a solution is obtained by imposing compatibility between the displacements of the pile and the adjacent soil for each element of the pile. The displacements of the pile are obtained by considering the compressibility of the pile under axial loading, while the soil displacements are obtained by using Mindlin's equations to determine the displacements within a soil mass due to loading within the mass.

Ideally, compatibility between both vertical and radial soil and pile displacements should be ensured to obtain a complete solution, but consideration of radial displacements involves considerable computational effort. Furthermore it has been shown by Mattes (1969) and Butterfield and Banerjee (1970), (1971) that the solution for displacement and shear stress along the pile are not significantly affected by considering only compatibility of the *vertical* displacements of the soil and pile.

The simplest case to analyse is that of a single floating pile in a uniform semi-infinite soil mass. However, to obtain solutions for more realistic problems, modifications may readily be made to this analysis for factors such as pile-soil slip, a pile resting on a stiffer bearing stratum, a pile with an enlarged base, piles of non-uniform diameter and piles having a cap resting on the surface. Moreover, the single pile analysis may readily be extended to the analysis of pile groups (Poulos (1968), Poulos and Mattes (1971a)) and to the calculation of settlements within a soil mass due to loaded piles or pile groups (Poulos and Mattes (1971)).

Step integration method.   This method was proposed by Coyle and Reese (1966) and utilises soil data measured from field tests on instrumented piles and laboratory tests on model piles. The relevant soil data required in this method are curves relating the ratio of the adhesion (or load transfer) and the soil shear strength, to the pile movement at various points along the pile. Such load transfer curves were first developed by Seed and Reese (1957). The basis of the method is to divide the pile into a number of elements, assume a tip movement, and from the load transfer curves, to determine the load and movement of successive elements up the pile until a value of load and displacement for the top of the pile are obtained. The procedure is repeated using different assumed tip movements, and a load-settlement curve may thus be computed.

Finite element method.   The development of the finite element method has provided a powerful tool for analysing pile behaviour. Non-homogeneous, anisotropic, non-linear and time-dependent stress-strain soil behaviour can be readily taken into account, and, special elements may also be used to allow for the possibility of pile-soil slip along the shaft, e.g. as used by Ellison *et al.* (1971) to analyse the behaviour of bored piles in clay.

Discussion of approaches. The finite element method is inherently the most powerful, but it requires a complete solution for each new problem considered. While a single pile is readily analysed as a radially-symmetrical problem, a pile group presents difficulties as the loss of axial symmetry requires that a three-dimensional analysis be carried out.

The step-integration method, although it has gained some acceptance, has a theoretical limitation in that it assumes that the movement of the pile at any point is related only to the shear stress at that point and is independent of the stresses elsewhere along the pile. This inherent assumption is the same as that made in the theory of subgrade reaction and hence no account is taken of the continuity of the soil mass. Consequently, this method cannot be used for analysing the load-settlement characteristics of pile groups.

The methods employing Mindlin's equation rely on the assumption of the soil as a linear elastic material although more realistic soil behaviour can be incorporated into the analysis in an approximate fashion. However, these methods provide a relatively rapid means of carrying out parametric studies of the effects of pile and soil characteristics, and of preparing a series of solutions which may be used for design without recourse to a computer. Furthermore, they may be used to develop a simple method of analysing the behaviour of pile groups; in this regard, the elastic approach has a considerable advantage over the other two methods considered.

In the following section, some of the recent solutions from the elastic analysis are summarised and presented in graphical form for easy design application. A simple practical method of utilising these solutions for estimating load-settlement behaviour to failure of a single pile, pier or pile group is also outlined.

7.3 SOLUTIONS FROM ELASTIC THEORY

7.3.1 Single piles

The more useful solutions have been summarised by Poulos (1972a), and the settlement at the top of the pile and the proportion of load transferred to the pile base have been expressed in terms of a basic influence factor for a rigid pile in a deep layer and correction factors for the effects of pile compressibility, finite soil layer depth and modulus of the bearing stratum at the base of the pile. For the two cases shown in *Figure 7.1*, the pile head settlement ρ is given by the following expressions:

(1) For a floating or friction pile in uniform soil,

$$\rho = \frac{PI_1}{E_s d}. R_K . R_h \tag{7.1}$$

(2) For a pile bearing on a stiffer stratum,

$$\rho = \frac{PI_1}{E_s d}. R_K . R_b \tag{7.2}$$

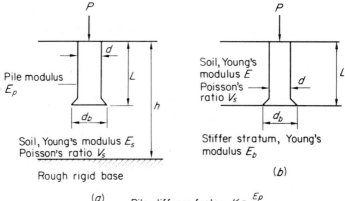

Figure 7.1   Definition of problem
(a) Floating or friction pile
(b) End-bearing pile

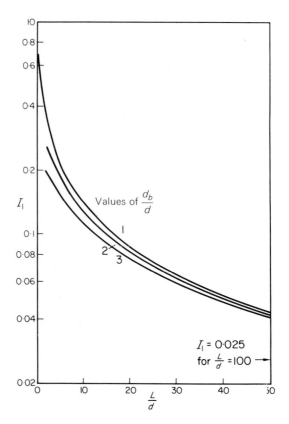

Figure 7.2   Displacement influence factor I_1

where P = applied load at pile head;

$\qquad d$ = pile diameter;

$\qquad E_s$ = Young's modulus of soil adjacent to shaft;

$\qquad I_1$ = settlement influence factor;

R_K, R_h, R_b = settlement correction factors for the effects of pile compressibility, finite layer depth and bearing stratum rigidity, respectively. A useful measure of the relative compressibility of the pile is the pile stiffness factor K, defined as

$$K = E_p/E_s \qquad (7.2a)$$

where E_p = effective Young's modulus of pile section.

Values of I_1, R_K, R_h and R_b are plotted in *Figures 7.2 to 7.5* as a function of the dimensionless factors L/d, d_b/d, K, h/L and E_b/E_s.

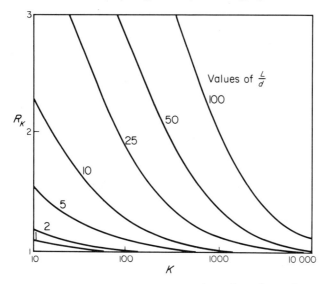

Figure 7.3   Compressibility correction factor for settlement R_K

The soil Poisson's ratio v_s has relatively little effect on the settlement factors. For $v_s = 0$, the value of I is generally reduced by about 15% compared with the value for $v_s = 0.5$, and this reduction varies almost linearly with increasing v_s.

Similarly, the proportion of load transferred to the base, β, is expressed as

$$\beta = \beta_1 C_K C_b \qquad (7.3)$$

where β_1 = value for a pile in a uniform deep soil layer;

$\qquad C_K$, C_b = correction factors for the effects of pile compressibility and bearing stratum rigidity.

For a floating pile, $C_b = 1$. The finite layer depth has virtually no effect. Values of β, C_K and C_b are plotted in *Figures 7.6 to 7.8*.

242

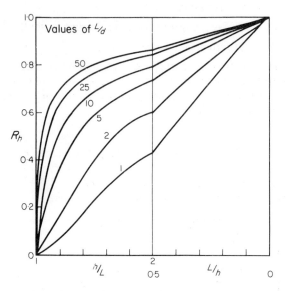

Figure 7.4   Depth correction factor for settlement R_h

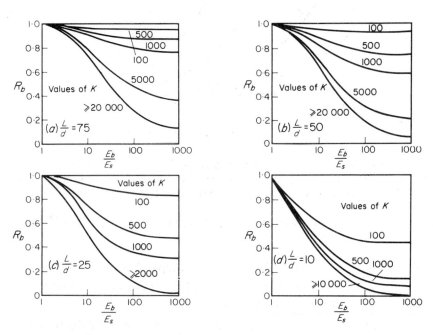

Figure 7.5   Base modulus correction factor for settlement R_b

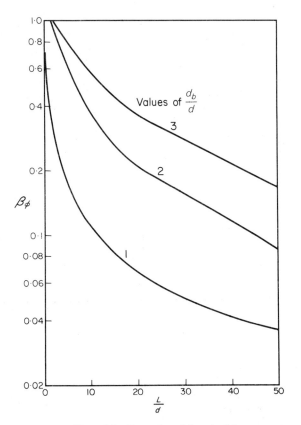

Figure 7.6   Proportion of base load β_ϕ

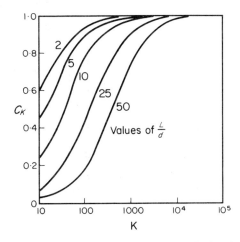

Figure 7.7   Compressibility correction factor for base load C_K

244

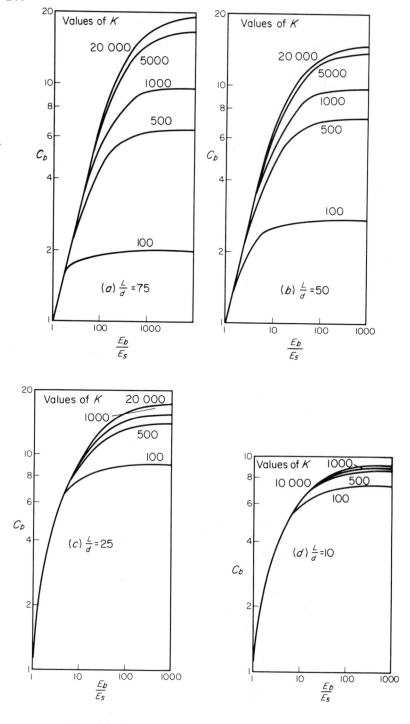

Figure 7.8   Base modulus correction factor for base load C_b

The solutions in *Figures 7.2* to *7.8* are valid only if the soil along the pile and at the pile base remain elastic. If pile-soil slip or base failure occurs, these elastic solutions may be utilised to construct the load-settlement curve to failure.

One of the significant conclusions arising from the elastic solutions is that most of the settlement of a pile occurs as immediate settlement and that only a relatively small proportion of the final settlement (generally between 5 and 20%) can be attributed to consolidation. This result may be derived by assuming the soil to be an ideal two-phase elastic material. The immediate settlement ρ_i is then calculated using the undrained Young's modulus E_u of the soil and the settlement influence factor I_1 for the undrained Poisson's ratio v_u ($v_u = 0.5$ for a saturated soil). The total final settlement is calculated using the drained Young's modulus E_s' of the soil and the value I_1 for the drained Poisson's ratio v_s'. For an ideal two-phase soil, $E_u = 3E_s'/2(1 + v_s')$, so that the ratio of immediate to total final settlement may be determined. The above theoretical conclusions regarding the importance of immediate settlement are supported by a considerable amount of field data from maintained loading tests.

7.3.2 Pile groups

From an analysis of two identical loaded piles within an elastic mass, solutions may be obtained for the increase in settlement of a pile due to the presence of the adjacent pile as a ratio of the settlement of a single isolated pile. This ratio, termed an 'interaction factor', is a function of the pile spacing, the pile stiffness factor K, L/d, the relative depth of the soil layer h/L, and the nature of the bearing stratum. Solutions for interaction factors have been presented by Poulos (1968) and Poulos and Mattes (1971a). By employing superposition of these interaction factors, it is possible to analyse the settlement and load distribution within any pile group.

From a practical point of view, the most useful output from such an analysis is a factor termed the 'group reduction factor', R_G, which expresses the decrease in settlement obtained by using a pile group rather than a single pile to resist a given load. R_G is simply related to the more familiar group settlement ratio R_s (the ratio of the group settlement to the settlement of a single pile under the same *average* load) as

$$R_G = R_s/n \tag{7.4}$$

where n = number of piles in the group.

The settlement of a group ρ_G is then given by

$$\rho_G = nR_G\rho_{1\,\mathrm{av}} \tag{7.5}$$

where $\rho_{1\,\mathrm{av}}$ = the single pile settlement under the average pile load.

$\rho_{1\,\mathrm{av}}$ may be obtained from a pile load test, or alternatively, calculated from the theoretical solutions in the previous section. If the latter alternative is employed, then from equations 7.1 and 7.2,

$$\rho_G = \frac{P_G I}{E_s d} R_G \tag{7.6}$$

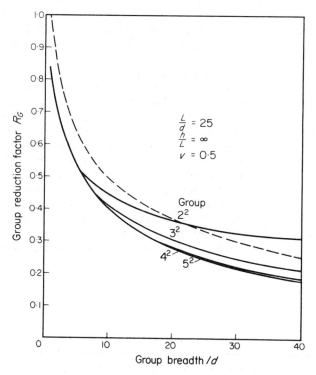

Figure 7.9 *Settlement against breadth of group incompressible piles with rigid pile cap*

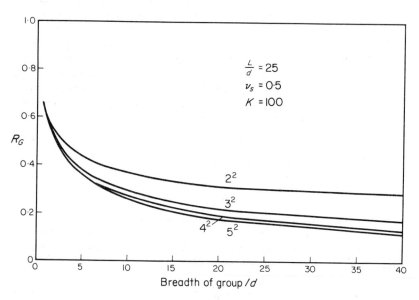

Figure 7.10 R_G *versus group breadth–floating groups*

where P_G = total load on group;

R_G = group reduction factor;

$I = I_1 R_K R_h$ for floating piles or $I_1 R_K R_b$ for end-bearing piles.

It has been found that, for a given set of pile characteristics, R_G is largely unaffected by the number or arrangement of piles in the group but is primarily a function of the breadth of the group. For groups containing more than 25 piles, it appears that, for a practical range of group breadths, a common limiting curve of R_G versus breadth, almost coincident with the

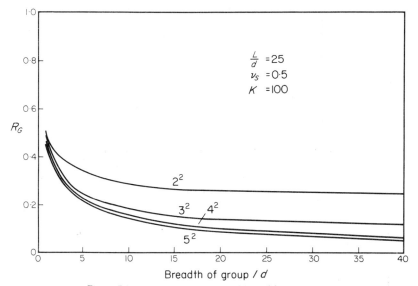

Figure 7.11 R_G versus group breadth–end-bearing groups

curve for a square 25 pile group, is obtained. For square groups of piles (denoted as 2^2, 3^2 etc.), typical plots of R_G versus dimensionless group breadth are shown in *Figures 7.9* and *7.10* for floating pile groups in a semi-infinite mass and in *Figure 7.11* for end-bearing groups resting on a rigid stratum. *Figures 7.12* and *7.13* give indications of the effect of L/d and K on R_G.

7.3.3 Simplified method for constructing load-settlement curve to failure

Analyses taking account of pile-soil slip along the shaft have revealed that for normal piles having length to diameter ratio, L/d, greater than about 20, the load-settlement curve is substantially linear until a load of at least 50–70% of the failure load is reached. For the prediction of the settlement at working loads for such piles, a linear elastic analysis is therefore adequate. However, for large diameter piles or piers, piles with an enlarged base or some pile groups, full shaft slip may occur at relatively low loads, so that some account should be taken of the effects of shaft slip on load-settlement behaviour.

248

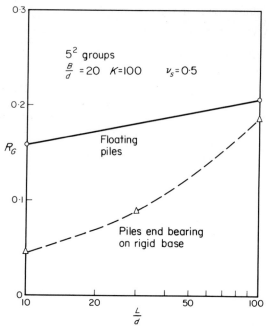

Figure 7.12   Effect of L/d on R_G

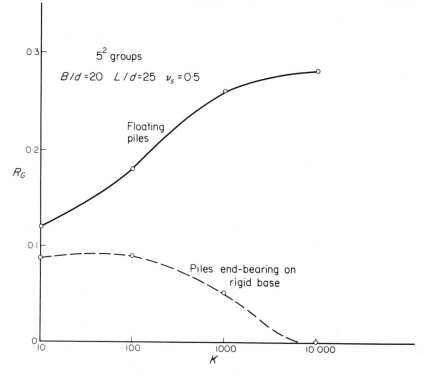

Figure 7.13   Effect of K on R_G

A relatively simple approach for constructing the load-settlement curve to failure for a pile or pier has been suggested by Poulos (1972a) in which the foregoing elastic solutions are utilised. The overall load-settlement curve is constructed as a combination of the relationships between shaft load *vs.* head settlement and base load *vs.* head settlement. These relationships are assumed to be linear up to failure of the shaft and base respectively. Justification for this simple assumption is provided by the tests of Whitaker and Cooke (1966), Williams and Colman (1965) and McCammon and Golder (1970). The procedure for the estimation of load *vs.* immediate settlement of

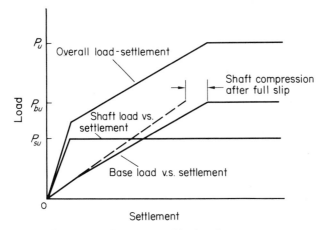

Figure 7.14   *Construction of load-settlement curve*

piles in clay or load *vs.* total settlement of piles in sand is shown in *Figure 7.14*. The shaft load *vs.* settlement relationship, up to the ultimate resistance P_{su} of the shaft, is

$$\rho = \frac{I}{E_s d} \cdot \frac{P_s}{(1-\beta)} \qquad (7.7)$$

where P_s = shaft load;

$I = I_1 R_K R_h$ for a floating pile or $I_1 R_K R_b$ for an 'end-bearing' pile.

The base load *vs.* settlement relationship, up to the ultimate resistance, P_{bu}, of the base is

$$\rho = \frac{I}{E_s d} \cdot \frac{P_b}{\beta} + \left(P_b - \frac{P_{su}\beta}{1-\beta}\right) \cdot \frac{L}{A_p E_p} \qquad (7.8)$$

where P_b = base load;

A_p = area of pile section;

E_p = Young's modulus of pile;

I = as in equation 7.7.

The second term in the above equation represents the additional compression of the shaft after full shaft slip and is only operative if the shaft has fully slipped.

The ultimate base and shaft loads P_{bu} and P_{su} may be estimated from ultimate load capacity theory; a review of various theories is given by Coyle and Sulaiman (1970).

For piles or piers in clay, consolidation settlements will occur and Poulos (1972a) has suggested that the consolidation process be considered as entirely elastic, even if some shaft slip has occurred under undrained conditions. The consolidation settlement at any load is therefore calculated as the difference between the total final and the undrained settlements, both calculated on a purely elastic basis (no yield).

Pile groups. To obtain a simple approximate solution for the load-settlement behaviour of a pile group, the ultimate group load may be first estimated as the lesser of the values to cause individual failure of the piles in the group, or failure of the group as a single block. If individual pile failure is found to be critical, i.e. the group efficiency is unity, the load-settlement curve for a single pile may be used except that, at any point, the single pile settlements are multiplied by the group settlement ratio R_s while the load is multiplied by the number of piles in the group.

If block failure is likely to occur, the simplest procedure is to replace the group by an equivalent single pier of equal gross area and an equivalent length (Poulos (1968)), and then compute the load-settlement curve for this equivalent pier.

7.3.4   Soil modulus E_s

One of the principal problems in applying elastic theory in soil mechanics is to determine an appropriate Young's Modulus for the soil, E_s. For pile foundations, E_s determined from conventional laboratory triaxial tests has been found to over-predict settlements by up to a factor of 10.

The most satisfactory means of obtaining E_s at present is to carry out a pile loading test *in situ* and to back-figure the average value of E_s from the measured settlements. In order to illustrate this procedure, an example is given later in this section. In relation to piers or caissons, a test on a full-scale pier may not be feasible unless it is carried out on a finished pier. As an alternative, a pile or pier of similar length but smaller diameter, installed in a similar manner to the final pier, may be tested. In relation to pile groups, provided that the test pile is of the same dimensions as the group piles, it may be adequate to measure the settlement of the test pile at the proposed design load and then to apply the theoretical value of settlement ratio R_s to predict the group settlement.

Table 7.1   AVERAGE VALUES
OF E_s FOR DRIVEN PILES
IN SAND

Sand density	Range of E_s ($kN.m^{-2}$)
Loose	27 500– 55 000
Medium	55 000– 69 000
Dense	69 000–110 000

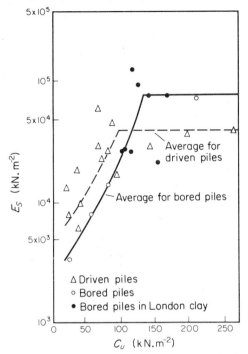

Figure 7.15   Backfigured soil modulus E_s for piles in clay

In situations where a load test cannot be performed or a rough estimate of settlement only is required, an alternative source of data is empirical values of E_s derived from published pile load test results. Relationships between E_s and undrained shear strength c_u, for piles in clay, are shown in *Figure 7.15*, while average values of E_s for piles in sand are shown in *Table 7.1*.

Values of drained Poisson's ratio v'_s generally range between 0.3 and 0.4 for most soils.

Example of interpretation of pile load test to backfigure E_s.   As a simple example, the case of 0.3 m dia 15 m long floating test concrete pile in a 30 m thick layer of clay will be considered. It will be assumed that, at a load of 9.8 kN, a settlement of 10 mm is recorded. In order to backfigure the average value of E_s of the clay, use is made of equation 7.1.

From *Figure 7.2*,

$$I_1 = 0.044 \quad \text{for} \quad L/d = 15/0.3 = 45 \quad \text{and} \quad d_b/d = 1$$

and $R_h = 0.87$ from *Figure 7.4*,   for $L/d = 50$ and $h/L = \frac{30}{15} = 2$

Substituting for ρ, P, d, I_1 and R_h into equation 7.1, it is found that

$$E_s = 6.26 \times 10^3 R_K \text{ kN.m}^{-2} \tag{7.9}$$

R_K is a function of pile stiffness factor K, and hence E_s. Thus, equation 7.9 must be solved together with the equation defining K, i.e.

$$K = E_p/E_s \tag{7.10}$$

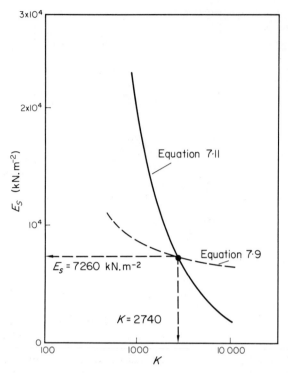

Figure 7.16   Graphical solution for determination of E_s from pile load test

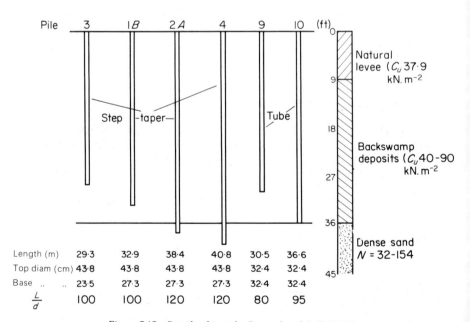

Figure 7.17   Details of tests by Darragh and Bell (1969)

Assuming $E_p = 196 \times 10^5 \, \text{kN.m}^{-2}$,

$$E_s = \frac{196 \times 10^5}{K} \, \text{kN.m}^{-2} \qquad (7.11)$$

Solution of equations 7.9 and 7.11 is most easily carried out graphically, using a tabulation such as is given in *Table 7.2*. The graphical solution is shown in *Figure 7.16* from which the required value of E_s is found to be 7260 kN.m^{-2}.

Table 7.2

K	R_K (*Figure 7.3*)	E_s (*equation 7.9*) ($kN.m^{-2}$)	E_s (*equation 7.11*) ($kN.m^{-2}$)
10 000	1.02	6370	1960
5000	1.08	6760	3920
2000	1.19	7440	9800
1000	1.37	8560	19 600
500	1.68	10 500	39 200

7.4   APPLICATIONS OF PILE SETTLEMENT THEORY TO MODEL AND FIELD TESTS

Although the majority of published pile load test results are concerned primarily with ultimate load capacity, some tests are sufficiently well-documented to allow the settlement behaviour to be analysed and compared with that predicted from pile settlement theory.

The cases chosen for consideration have been divided into two categories, those involving single piles only and those concerned with pile groups. Most cases involve field tests although some model-scale laboratory tests are also included.

7.4.1   Settlement behaviour of single piles

Tests by Darragh and Bell (1969).   Mattes (1972) has analysed an interesting series of pile tests carried out by Darragh and Bell (1969) at the site of Gulf Oil Corporation's Faustina Works, on the banks of the Mississippi River. Brief details of the piles driven and of the site subsurface conditions are given in *Figure 7.17*.

The site involved about 36.6 m of natural levee and back-swamp deposits consisting of layers and laminations of clays, silts, and fine sands, which overlay a 21 m deep layer of fine silt grading to sandy gravel with depth. Two pairs of step-taper piles and one pair of steel tube piles were driven; step-taper piles 1B and 3 were friction piles driven to within 6 m of the sandy stratum, step-taper piles 2A and 4 were end-bearing in the sandy stratum, tube pile 9 was a floating pile founded at a similar depth to 1B and 3, and tube pile 10 was end-bearing in the sand. Piles 1B and 3 gave a very similar load-test result, and were analysed as floating piles in a finite layer to derive a backfigured soil modulus for the back-swamp deposits. The relevant details of these piles are as follows:

Pile length, L	32.9 m
Average diameter, d	0.33 m
L/d	100 (approx)
Depth of founding layer, h	36.6 m
h/L	1.1 (approx)
Ultimate Load	1600 kN
Settlement at 800 kN	3.04 cm
Soil Modulus E_s (back-figured)	44 800 kN.m^{-2}
Pile stiffness factor, K	460–500

For the purpose of predicting the performance of the end-bearing piles 2A, 4 and 10, it was assumed that the base soil modular ratio, E_b/E_s, was 2.

In *Table 7.3* the observed settlements of piles 2, 4A, 9 and 10 are compared with those predicted by the elastic theory, using the soil modulus derived from the floating pile tests 1B, and 3, and it can be seen that the settlement performance of both the floating and end-bearing piles has been closely predicted.

Large bored piers. For some of the tests reported by Whitaker and Cooke (1966) and Burland *et al.* (1966) on bored piers in London Clay, comparisons are shown in *Figures 7.18 to 7.21* between measured and predicted shaft load *vs.* settlement, and base load *vs.* settlement. The values

Table 7.3 PREDICTED PILE PERFORMANCE—TESTS BY DARRAGH and BELL (1969)

Pile No.	2A	4	9	10
Floating or end-bearing	End-bearing	End-bearing	Floating	End-bearing
Length (m)	38.4	40.2	30.4	36.6
Length/diameter (approx)	120	120	80	95
Pile type	Steel step-taper (closed)	Steel step-taper (closed)	Closed steel tube 0.188 in wall	Closed steel tube 0.188 in wall
E_s (kN.m^{-2})	44 700	44 700	44 700	44 700
E_b/E_s	2	2	1	2
K	500	500	270	270
Applied load (kN)	1200	1200	400	800
Predicted top settlement (cm)	0.64	0.64	0.23	0.41
Measured top settlement (cm)	0.51	0.61	0.25	0.43
Predicted base settlement (cm)	0.13	0.13	0.03	0.04
Measured base settlement (cm)	0.10	0.15	0.04	0.05

of E_s in these comparisons have been obtained from the mean curve in *Figure 7.15*. The agreement is generally reasonable and the linearity assumption of these relationships appears justified, at least over the range of load employed in these tests. Comparisons between measured and predicted load-settlement behaviour are shown in *Figures 7.20* and *7.21*.

Reasonable agreement is again found and at least part of the discrepancy which exists can be attributed to the selection of E_s from the average curve in *Figure 7.15* rather than using the actual back-calculated value of E_s for the particular test. Also shown in *Figure 7.20*, for two of the piles are the ones

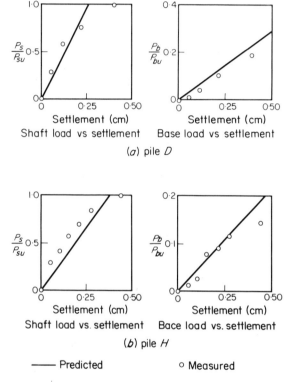

Figure 7.18   Comparisons between predicted and observed behaviour. Tests by Whitaker and Cooke (1966)

predicted by Ellison *et al.* (1971) from a finite element analysis using a tri-linear stress-strain curve for the London Clay. These predicted curves agree well with the measured curves and are in reasonable agreement with those predicted by the approximate approach described herein.

Tests by Mansur and Kaufman (1956).   Six instrumented piles were driven into a fairly deep layered system of silts, sandy silts, and silty sands with interspersed clay strata, underlain by a deep layer of dense fine sand. All except pile number 5 were driven to end-bearing in the dense fine sand; pile 5 was a floating pile. One of the end-bearing piles (pile 3) was an *H*-pile with a rectangular base plate attached, and because of the disturbing effect

256

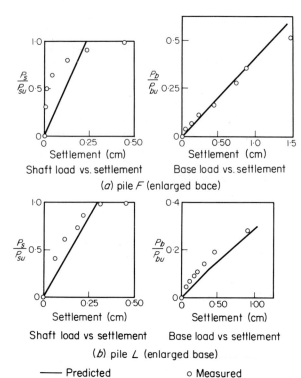

Figure 7.19 *Comparisons between predicted and observed behaviour. Tests by Whitaker and Cooke (1966)*

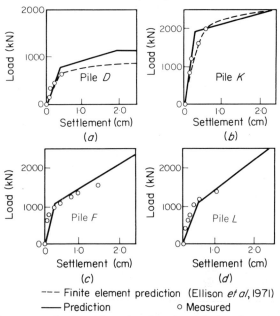

Figure 7.20 *Comparisons between observed and predicted load-settlement curves. Tests of Whitaker and Cooke (1966)*

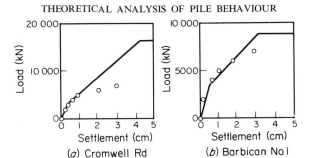

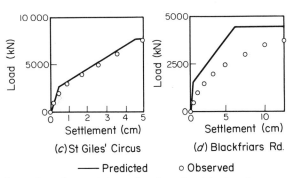

── Predicted   ○ Observed

Figure 7.21   Comparisons between observed and predicted load-settlement curves. Tests of Burland et al. (1966)

of this plate during driving, was not considered. The test results were analysed as follows:

(1) Using the single pile theory, pile 5 was analysed as a floating pile in a finite layer, and a soil modulus, E_s, of 6.9×10^4 kN.m^{-2} was back-figured from the pile test results.
(2) From the SPT blow counts for the silty soils and the dense fine sand, it was deduced that a ratio of soil to bearing stratum moduli, E_b/E_s, of about 3 would be applicable for the end-bearing piles.
(3) The analysis for a single end-bearing pile Poulos and Mattes (1969), was used to evaluate the load-distribution along, and settlement of, piles 1, 2, 4 and 6. In *Table 7.4*, the details of pile properties, settlements and settlement predictions are given, while in *Figures 7.22a to e*, the predicted and measured load distributions along the piles are compared.

Figure 7.22 and *Table 7.4* show that quite low values of K are possible when steel tubes or H-sections are used as piles. In such cases, it is likely that very little load does in fact reach the pile base, even in nominally end-bearing piles. In the case described here, the results of a floating pile test, when combined with the results of a routine borehole test, have allowed the accurate prediction of the load distribution along, and settlement of, end-bearing piles on the same site.

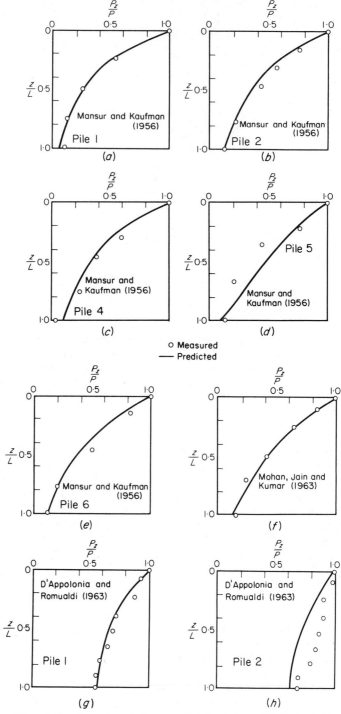

Figure 7.22 Comparisons between predicted and observed load distributions

Table 7.4 TESTS BY MANSUR and KAUFMAN (1956)

Pile No.	1	2	4	5	6
Pile type	0.36 m H-beam	0.53 m pipe	0.43 m pipe	0.43 m pipe	0.48 m pipe
Length (m)	24.7	19.8	20.1	13.7	19.8
L/d	70	37	47	32	41
K	470	250	350	350	350
End-bearing (EB) or floating (F)	EB	EB	EB	F	EB
E_b/E_s	3	3	3	(1)	3
Load (kN)	1250	1250	1250	750	750
Observed settlement (cm)	0.33	0.33	0.41	0.25	0.33
Predicted settlement (cm)	0.37	0.33	0.36	(0.25)	0.38
Predicted/Observed settlement	1.10	1.00	0.89	(1.00)	1.15

Note: Soil type: alternating strata of silts, silty sands, sandy silts, with interspersed clay strata.
Bearing stratum: dense fine sand.
Pile 5 (floating) used as control pile for predictions.

Test by Mohan, Jain and Kumar (1963). A 0.36 m diameter cast *in situ* pile having $L/d = 33$ was placed through a layered system of fill, medium sand and silt to end-bearing on a bed of dense fine sand. No satisfactory soil data was available in the test report, and so a value of K of 300 was assumed, based on an E_s value from *Table 7.1*, and a ratio of bearing stratum to soil modulus of 2 was adopted. The observed and calculated load distributions are shown in *Figure 7.22f*, and reveal good agreement.

Tests by D'Appolonia and Romualdi (1963). Tests on two instrumented H-piles were reported; the piles were about 13.7 m long, and passed through layers of fill, sandy silt, sand and gravel, fine to medium sand, sand and gravel, and sandy silt, to end-bearing in shale. No satisfactory soil data was available, so a K value for solid steel piles of 3000 was adopted, based on an E_s value from *Table 7.1*. The bearing stratum was assumed to be rigid. In *Table 7.5*, the pile properties and settlement details are listed, and comparisons based on the assumed soil properties are made. In *Figures 7.22g and h*, the load distributions within the piles are compared with the calculated distributions. In each case, quite reasonable agreement between prediction and observation is obtained.

Model tests. A series of carefully controlled model tests on piles in normally consolidated clay has been carried out by Mattes and Poulos (1971) in order to examine the effects of length-to-diameter ratio and pile compressibility on settlement, and the relative proportions of immediate and final settlement. *Figure 7.23* shows the ratio of predicted to observed settlements for piles of various L/d, using brass piles having $L/d = 25$ as control piles and back-figuring the undrained and drained Young's moduli E_{su} and E_s', from the measured settlements of these piles. The agreement between

Table 7.5   TESTS BY D'APPOLONIA and ROMUALDI
(1963)

Pile No.	1	2
Pile type	14 BP 89 (H-pile)	14 BP 119 (H-pile)
Length (m)	13.4	13.7
Assumed L/d	33	34
Assumed area ratio (R_A)	0.143	0.186
K	430	560
E_b/E_s	—	—
Load (kN)	750	1000
Observed settlement (cm)	0.18	0.28
Predicted settlement (cm)	0.15	0.23
Predicted/Observed settlement	0.86	0.82

Note: Soil: layers of fill, sandy silt, sand and gravel, fine to medium sand, sand and
gravel, and sandy silt.
Bearing stratum: shale.

predicted and measured settlements for $L/d = 10$ and 40 is reasonably good, indicating that the theory predicts with adequate accuracy the effects of pile length on settlement. Also shown are comparisons for plastic piles, of about 1/10th of the stiffness of the brass piles, having $L/d = 40$. Again the good agreement indicates that the theory gives a good prediction of the effects of pile compressibility.

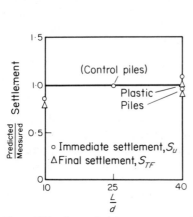

Figure 7.23   Comparison between measured and predicted settlements (after Mattes and Poulos (1971))

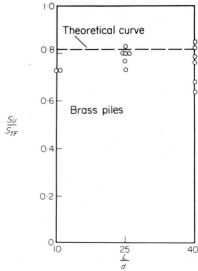

Figure 7.24   Comparison between observed and predicted ratio S_u/S_{TF} (after Mattes and Poulos (1971))

Figure 7.24 shows a comparison between measured and predicted ratios of immediate to final settlement of the model piles, and reveals fair agreement. The test results confirm the conclusion reached from the theory that the major part of the settlement of a pile occurs as immediate settlement and that consolidation settlement is relatively unimportant for ordinary piles at normal working loads.

7.4.2 Observed and theoretical group behaviour—settlements

A number of comparisons between measured and theoretical values of settlement ratio for floating pile groups were made by Poulos and Mattes (1971a). A summary of the cases considered is given in *Table 7.6* and the comparisons are shown in *Figure 7.25*. In all cases, the load level corresponds to a factor of safety of at least 2 against ultimate failure of the group. With the exception of the model tests by Hanna (1963) in loose sand, the agreement is generally satisfactory for both large and small values of *K*. The poor

Table 7.6 SUMMARY OF DATA ON FLOATING PILE GROUP TESTS

Test	Pile material	Soil type	Assumed parameters for comparisons			Remarks
			L/d	K	Layer depth/L	
Whitaker (1957)	Brass	Remoulded London clay	24	∞	2	Model tests
Saffery and Tate (1961)	Stainless steel	Remoulded clay	20	∞	2	Model tests
Sowers *et al* (1961)	Aluminium tube	Remoulded bentonite	24	2000	2	Model tests
Berezantzev *et al* (1961)	Concrete	Dense sand	20	1000	∞	Field tests K estimated for quoted values of E_s
Hanna (1963)	Wood	Dense sand	33	100	2	Model tests
Hanna (1963)	Wood	Loose sand	33	1000	2	

agreement for the tests in loose sand may be attributed to the effects of the greater densification of the loose sand by the pile group as compared with the single pile. These comparisons therefore indicate that the theoretical approach should be satisfactory in practical cases except for pile groups in loose sand.

Poulos (1972b) presented a comparison of the predicted and observed relationship between settlement and number of piles for groups of end-bearing piles in Boston Blue clay, using measurements reported by D'Appolonia and Lambe (1971) for four buildings on the MIT campus. This comparison is shown in *Figure 7.26* and shows good agreement between the predicted relationship and the regression line for the measured relationship.

Group behaviour predicted from single pile test results. Tests by Koizumi and Ito (1967). A series of full scale tests by Koizumi and Ito (1967) was studied by Mattes (1972) in an attempt to predict the performance of a pile group from the results of a single pile test. A single floating pile and a

262

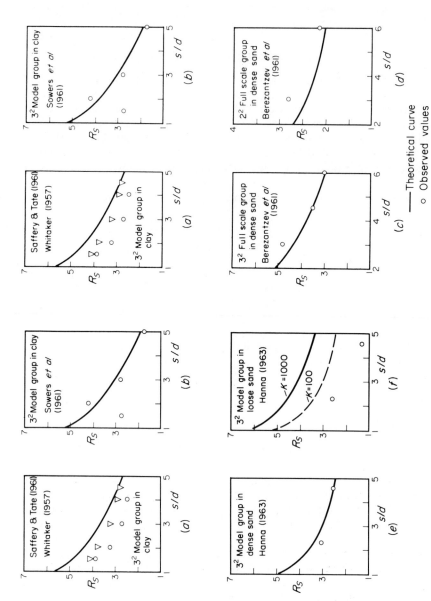

Figure 7.25   Comparison between theoretical and observed settlement ratios

nine-pile rigid-capped group of similar piles were founded in a thick uniform layer of silty clay overlain by a thin layer of sandy silt. The piles were closed-end steel tubes, and were instrumented to allow pile loads, earth pressures and pore pressures to be measured. Provision was also made for measuring displacements and pressures in the soil remote from the piles. Details of the foundations and site conditions are given in *Figure 7.27*.

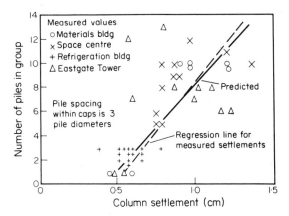

Figure 7.26 *Comparison between measured and predicted group settlements*

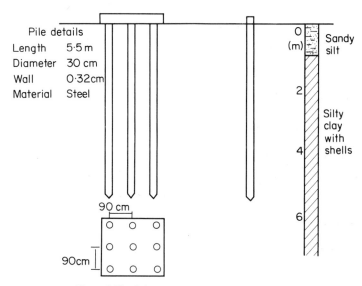

Figure 7.27 *Pile group test (Koizumi and Ito)*

By using the single pile load test results, a soil modulus of 17 250 kN.m^{-2} was back-figured, corresponding to a pile stiffness factor of 500, at a load factor against failure of approximately 2.5. From pile group theory the settlement of the rigid-capped group (at the same load factor) has been calculated and compared with the measured settlement, as follows:

Group	3^2, floating, rigid cap.
L/d	18.5 (individual piles).
Spacing	3 pile diameters centre-to-centre.
Group load, P_G	91 000 kg
Group reduction factor R_G	0.40
Single pile settlement for unit load, ρ_1	0.2×10^{-2} mm.kN^2
Predicted settlement of group $P_i \rho_1 R_G$	7.2 mm
Measured settlement of group	7.1 mm

Although the cap of this group was in contact with the soil the effect of the cap on group settlement is negligible (Davis and Poulos (1972)) and it will be seen that there is excellent agreement between predicted and measured group settlement. The measured and theoretically predicted load distributions within the group also agree closely, as is shown in *Table 7.7*

Table 7.7   THEORETICAL AND MEASURED LOAD DISTRIBUTION TESTS OF KOIZUMI and ITO (1967)

Pile location	Pile load/Average pile load	
	Theoretical	Measured§
Centre	0.35	0.46
Corner	0.82	0.86
Mid-side	1.35	1.20

§Group load 1200 kN.

Soil displacements near the group were recorded mainly in connection with ultimate bearing capacity investigations, and at working loads are considerably smaller than those predicted theoretically.

Model group tests by Mattes and Poulos (1971).   Tests were carried out on 3 × 3 and 6 × 1 floating pile groups with piles of $L/d = 25$, at a pile spacing of two diameters. The measured settlements are compared in *Table 7.8* with those predicted from the results of single pile tests carried out under the same conditions, and it is seen that there is close agreement between observed and predicted settlements.

7.5   WAVE EQUATION ANALYSIS OF PILE DRIVING

7.5.1   Introduction

The earliest and perhaps still the most frequently used method of estimating the load capacity of driven piles is to use a pile driving formula. Such a formula relates ultimate load capacity to pile set by using an individual

Table 7.8 MODEL PILE GROUP TESTS (MATTES and POULOS)

Group size		3×3	6×1
Total group load (N)		223	180
Theoretical group settlement ratio R_s		4.77	3.24
Settlement of single pile at average pile load (predicted from average soil modulus backfigured from single pile tests) (cm $\times 10^{-4}$)	Immediate	9.7	11.2
	Final	12.6	14.1
Immediate settlement of group (cm $\times 10^{-4}$)	Predicted	46.2	36.2
	Observed	38.8	38.1
	Ratio of predicted to observed settlement	1.19	0.95
Final settlement of group (cm $\times 10^{-4}$)	Predicted	59.4	45.7
	Observed	61.4	50.8
	Ratio of predicted to observed settlement	0.98	0.90

representation of the action of the hammer on the pile in the last stage of its embedment, assuming that the driving resistance is equal to the load capacity of the pile under static loading.

A relatively recent improvement in the estimation of load capacity by dynamic methods has been the use of the wave equation to examine the transmission of compression waves down the pile, rather than using rigid body mechanics and assuming that a force is generated instantly throughout the pile, as is done in deriving driving formulae. The main objective in using the wave equation approach is to obtain a better relationship between ultimate pile load and pile set than can be obtained from a simple driving formula. However, the wave equation approach also enables a rational analysis to be made of the stresses in the pile during driving and can therefore be useful in the structural design of the pile.

The use of the wave equation was considered by Isaacs (1931) and Glanville et al. (1938), but it was not until the work of Smith (1960) that the method was fully developed. Considerable refinements to the analysis of Smith have been made, notably by Samson, Hirsch and Lowery (1963) and Forehand and Reese (1964). In the description of the method herein, the nomenclature and notation of Samson et al. (1963) and Hirsch et al. (1970) will generally be employed. The numerical solution of the wave equation to obtain the ultimate load versus set relationship will be described and typical solutions will be given. Some applications of the analysis to practical problems will be discussed and a comparison made between the reliability of this approach and some of the conventional pile driving formulae.

7.5.2 Analysis

For a prismatic bar embedded in a medium and subjected to an impact at one end, consideration of the internal forces and resulting motions yields the following form of the wave equation:

$$\frac{\partial^2 D}{\partial t^2} = \frac{E}{\rho}\frac{\partial^2 D}{\partial x^2} \pm R \tag{7.12}$$

where D = longitudinal displacement of a point of the bar from its original position;

E = modulus of elasticity of bar;

ρ = density of bar material;

t = time;

x = direction of longitudinal axis;

R = resistance of surrounding medium.

Because of the complications involved in practical piling problems, analytical solutions to equation 7.12 are not generally feasible and hence resort must be made to numerical means of solution. A convenient numerical method has been described by Smith (1960).

Smith's idealisation.   The method developed by Smith is a finite difference method in which the wave equation is used to determine the pile set for a given ultimate pile load. The pile system is idealised as shown in *Figure 7.28* and consists of

(1) A ram, to which an initial velocity is imparted by the pile driver.
(2) A capblock (cushioning material).
(3) A pile cap.
(4) A cushion block (cushioning material).
(5) The pile.
(6) The supporting soil.

The ram, capblock, pile cap, cushion block and pile are represented by appropriate discrete weights and springs. The frictional resistance on the side of the pile is represented by a system of springs and dashpots (*see Figure 7.30*) while the point resistance is represented by a single spring and dashpot. The characteristics of the components are considered subsequently. If the actual situation differs from that shown in *Figure 7.28*, e.g. if the cushion block is not used or if an anvil is placed between the ram and capblock, the idealisation of the system can be readily modified.

Pile model—internal springs.   The ram, capblock, pile cap and cushion block may be considered to consist of internal springs although the ram and pile cap may often be treated as rigid bodies. The load-deformation behaviour of these elements is most simply taken to be linear (*Figure 7.29a*) although internal damping may also be considered, e.g. as shown in *Figure 7.29b*, for components such as the capblock and the cushion block. It should be noted that the spring $K(2)$ in *Figure 7.28* represents both the cushion block and the top element of the pile and its stiffness may be obtained from Kirchoff's equation as

$$\frac{1}{K(2)} = \frac{1}{K(2)_{\text{cushion}}} + \frac{1}{K(2)_{\text{pile}}}$$ (7.13)

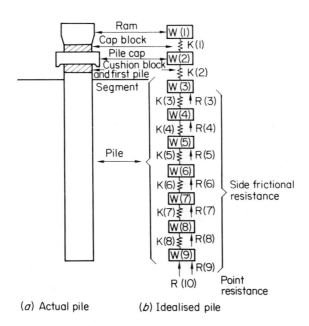

(a) Actual pile   (b) Idealised pile

Figure 7.28   Idealisation of pile

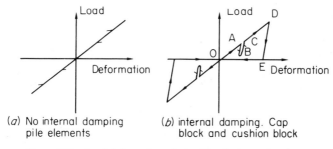

(a) No internal damping   (b) internal damping. Cap
pile elements   block and cushion block

Figure 7.29   Load-deformation relationships for internal springs

Soil model—external springs.   Smith's model of the load-deformation characteristics of the soil, represented as external springs subjected to static loading, is shown in *Figure 7.30a*. The path OABCDEFG represents loading and unloading in side friction. For the point, only compressive loading is

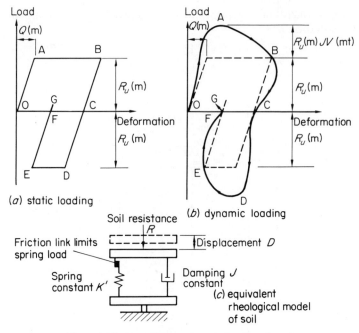

Figure 7.30   Load-deformation relationships for soil

considered and the loading and unloading path is OABCF. The quantities defining this static behaviour are Q and R_u, where

Q = 'quake', the maximum soil deformation which may occur elastically;

R_u = ultimate static soil resistance.

A load-deformation diagram such as *Figure 7.30a* may be established separately for each spring, so that

$$K'(m) = \frac{R_u(m)}{Q(m)} \tag{7.14}$$

where $K'(m)$ is the spring constant during elastic deformation for external spring m.

To allow for the effects of dynamic loading during driving in increasing the instantaneous resistance of the soil, the dynamic load-settlement behaviour of the soil is taken to be that shown in *Figure 7.30b*, which as pointed out by Hirsch *et al.* (1970), corresponds to a Kelvin rheological model. This dynamic behaviour is characterised by a further parameter J, the damping constant. The dashpot in the model produces an additional resisting force proportional to the velocity of loading V.

It has been shown by Smith (1960) that the wave equation may be expressed by a system of five simple equations as follows:

$$D(m, t) = D(m, t-1) + \Delta t \cdot V(m, t-1) \tag{7.15}$$

$$C(m, t) = D(m, t) - D(m+1, t) \tag{7.16}$$

$$F(m, t) = C(m, t) . K(m) \tag{7.17}$$

$$R(m, t) = [D(m, t) - D'(m, t)] . K'(m) . [1 + J(m) . V(m, t-1)] \tag{7.18}$$

$$V(m, t) = V(m, t-1) + [F(m-1, t) + W(m) - F(m, t) - R(m, t)] . \frac{g\Delta t}{W(m)} \tag{7.19}$$

where m = element number;

t = time;

Δt = time interval;

$C(m, t)$ = compression of internal spring, m, in time interval, t;

$D(m, t)$ = displacement of element, m, in time interval, t;

$D'(m, t)$ = plastic displacement of external spring, m, in time interval, t;

$F(m, t)$ = force in internal spring, m, in time interval, t;

g = acceleration due to gravity;

$J(m)$ = soil damping constant at element m;

$K(m)$ = spring constant for internal spring, m;

$K'(m)$ = spring constant for external spring, m;

$R(m, t)$ = force exerted by external spring m on element, m, in time interval, t;

$V(m, t)$ = velocity of element, m, at time, t;

$W(m)$ = weight of element, m.

Equation 7.17 applies for the elastic pile elements for which internal damping is ignored. For elements such as the capblock and cushion block, in which internal damping should be considered, the following equation should be used instead of equation 7.17:

$$F(m, t) = \frac{K(m)}{[e(m)]^2} . C(m, t) - \left\{ \frac{1}{[e(m)]^2} - 1 \right\} . K(m) . C(m, t)_{max} \tag{7.20}$$

where $e(m)$ = coefficient of restitution of internal spring, m;

$C(m, t)_{max}$ = temporary maximum value of $C(m, t)$.

The solution of the above equations proceeds as follows:

(1) A value of ultimate pile load R_u is assumed, from which the external (soil) spring constants may be computed, assuming a distribution of load along the shaft and at the point.
(2) The initial velocity of the ram $(VO(1,0)$ is determined from the properties of the pile driver. Other time-dependent quantities are initialised at zero or to satisfy static equilibrium conditions.
(3) Displacements $D(m, 1)$ are calculated from equation 7.15.
(4) Compressions $C(m, 1)$ are calculated from equation 7.16.
(5) Internal spring forces $F(m, 1)$ are calculated by equation 7.17.
(6) External spring forces $R(m, 1)$ are calculated by equation 7.18.
(7) Velocities $V(m, 1)$ are calculated by equation 7.19.

(8) Steps 2–7 are repeated for successive time intervals until the permanent set of the pile point $D'(p, t)$ reaches a maximum value.

(9) A new value of R_u is chosen and the procedure repeated to obtain the relationship between permanent set and R_u.

It is obvious that the above procedure requires the use of a computer for practical problems. A sample computer programme has been given by Bowles (1968).

In employing the numerical procedure described above, the accuracy of the resulting solution will depend on the values of time interval Δt and pile element length ΔL chosen. It has been shown that, for free longitudinal vibrations in a continuous elastic bar, the discrete element solution is an exact solution of the partial differential equation when

$$\Delta t = \frac{L}{\sqrt{(E/\rho)}} \qquad (7.21)$$

where L = length of pile.

Because inelastic springs and material of different densities and elastic moduli are usually involved in practical problems, Samson et al. (1963) recommended a value of Δt of about half the value given by equation 7.21. The accuracy of the solution is more sensitive to the choice of Δt if the pile is divided into only a few elements. The solutions of Samson et al. suggest that $\Delta L = L/10$ is generally a reasonable division of the pile. Smith's original suggestions for Δt and ΔL were $1/4000$ s and 2.5–3.0 m respectively for most practical piles, which are consistent with the values recommended above.

It must be emphasised that the above procedure, or indeed any dynamic method of load prediction, will only give the pile capacity immediately after driving. For piles in normally consolidated clay, the pile capacity tends to increase subsequent to driving (the phenomenon of 'set-up') and may reach a value two to three times the capacity just after driving, especially in soft soils. On the other hand, McClelland et al. (1969) point out that, for piles driven into stiff clay or sand, a decrease in load capacity with time may occur. An approximate modification to the predicted load capacity to allow for set-up effects is suggested by Lowery et al. (1969).

7.5.3 Parameters required in the analysis

Ultimate soil resistance, R_u.   Various values of R_u are input into the computer programme and the corresponding permanent set determined. The main problem with R_u is to determine the relative proportions of shaft and base resistance. A reasonable estimate of these proportions may be made by estimating the static shaft and base resistances from the known or assumed soil properties. A somewhat higher ultimate resistance for a given driving resistance is obtained if some shaft resistance is considered rather than only end-bearing. As a rough guide where other information is not available, values of the percentage of shaft resistance suggested by Forehand and Reese (1964) are shown in *Table 7.9.*

Quake Q.   Values of Q have been obtained empirically to date, and the single empirical values of Q for all elements of the pile suggested by Forehand

and Reese (1964) are shown in *Table 7.9*. It is however also possible to derive values of Q theoretically from pile settlement theory if the 'elastic' soil parameters are known (see Section 7.2). On the basis of this theory, the value of Q varies along the pile, with the value at the pile tip being greater than the values along the shaft. Alternatively, Q could also be estimated from the soil resistance curves employed by Seed and Reese (1957) and Coyle and Reese (1966). It is found that, for a given pile set, the value of R_u tends to decrease as the input value of Q increases.

Table 7.9 EMPIRICAL VALUES OF
Q, J AND % SIDE ADHESION
(after Forehand and Reese (1964))

Soil	Q (cm)	$J(p)$ (s/m)	Side adhesion (% of R_u)
Coarse sand	0.25	0.49	35
Sand gravel mixed	0.25	0.49	75–100
Fine sand	0.38	0.49	100
Sand and clay or loam, at least 50% of pile in sand	0.51	0.65	25
Silt and fine sand underlain by hard strata	0.51	0.65	40
Sand and gravel underlain by hard strata	0.38	0.49	25

Damping factor J. Empirical correlations between J and soil type have been obtained by Forehand and Reese (1964) and are shown in *Table 7.9*. The values in this table are for the pile point (i.e. $J(p)$). The average value for the sides of the pile $J(m)$ have been found to be less than $J(p)$ and for practical purposes, it has been suggested that

$$J(m) = 1/3 J(p) \qquad (7.22)$$

For a given pile set, it is found that R_u decreases as $J(p)$ increases.

One of the most significant parameters involved in pile driving is the energy output of the hammer, which must be known or assumed before the wave equation or dynamic formula can be applied. Measurements of energy output have been made recently (Housel (1966), Davisson and McDonald (1969)) and a summary of some available data is given by Hirsch et al. (1970). The ultimate resistance for a given set increases as the hammer energy increases.

Cap block and cushion data. The stiffness and coefficient of restitution of the cap block and cushion materials must be input into the analysis, and a summary of available data is given by Hirsch et al. (1970). A higher cushion stiffness results in an increase in pile resistance but also leads to higher driving stresses within the pile.

272

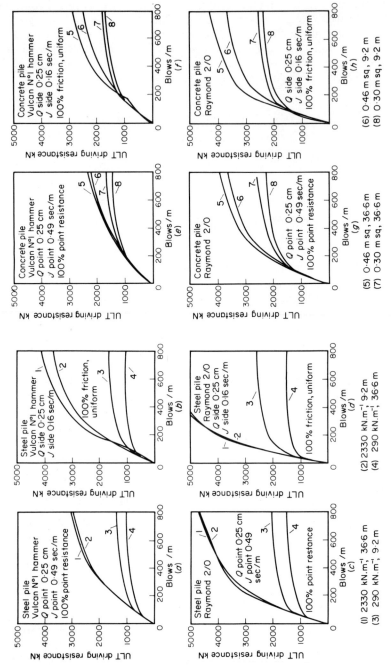

Figure 7.31   Solutions for the wave equation (after Mosley and Raamot (1970))

7.5.4 Typical solutions

A useful series of parametric solutions have been published by Mosley and Raamot (1970) for a range of commonly used steel and concrete piles driven for support of land structures. The parameters used are shown in *Table 7.10*.

The solutions are plotted in *Figure 7.31* as curves relating ultimate load capacity (R_u) to the number of blows per m, for assumed values of quake Q and damping factor J. The capblock is assumed to be micarta-aluminium and for the concrete piles, a 100 mm timber cushion is considered.

Table 7.10 PARAMETERS USED IN SOLUTIONS
OF MOSLEY and RAAMOT (1970)

Pile material	Steel, concrete	
Pile size	292,	3200
Steel $(N \cdot m^{-1})$		
Concrete		
0.46 m sq		2200
0.30 m sq		4900
Pile length (m)	9.15,	36.6
Mode of soil resistance %		
Point		100
Friction		100
Hammer size $(N \cdot m)$		
No. 1 Vulcan		20 400
No. 2/0 Raymond		44 100

Mosley and Raamot also compare the wave equation solutions with those from three well-known driving formulae, the Engineering News formula, the Eytelwein formula and the Hiley formula. The apparent factors of safety at the time of driving for the three formulae, based on the assumption that the wave equation solutions are correct, are given in *Table 7.11*. Two typical driving resistances, moderate (197 blows/m) and hard (790 blows/m) are considered. *Table 7.11* shows that there are certain combinations of parameters that cause the dynamic formulae to yield a desirable factor of safety, (e.g. 2), but for other combinations, these formulae may be either unsafe or else extremely conservative.

Pile stresses. For typical cases involving prestressed concrete piles, Samson *et al.* (1963) have investigated the effect of the Young's modulus of the pile and the stiffness of the cushion block on the maximum tensile and compressive stresses in the pile. Higher stresses occur for higher values of pile modulus or cushion block stiffness.

Especially significant, in relation to the design of prestressed concrete piles, were the high tensile stresses developed in the pile (of the order of $14\,000$–$28\,000$ kN $\cdot$ m^{-2} for the cases considered). It was noted that the time for a tensile wave to be reflected back along a typical pile is less than 0.02 s whereas a typical time interval between successive blows is of the order of 0.5 s. Thus, successive blows cannot be relied upon to reduce tensile stresses. It should be emphasised that conventional driving formulae predict that only compressive stresses occur within the pile during driving.

Table 7.11 RATIO OF WAVE EQUATION ULTIMATE RESISTANCE TO OTHER FORMULAS
(after Mosley and Raamot, 1970).

Pile type	Pile size ($N.m^{-1}$)	Pile length (m)	Vulcan 1				Raymond 2-0			
			Point		Friction		Point		Friction	
			197 Blows per m	790 Blows per m	197 Blows per m	790 Blows per m	197 Blows per m	790 Blows per m	197 Blows per m	790 Blows per m
Engineering-News Formula Safe Resistance										
Steel	291	9.15	1.9	1.4	2.4	1.7	1.5	1.0	2.1	1.2
Steel	291	36.6	1.4	1.0	1.8	1.1	1.0	0.7	1.3	0.7
Steel	2330	9.15	2.6	3.0	3.8	3.7	2.7	2.3	3.9	2.7
Steel	2330	36.6	2.6	3.0	4.1	4.1	2.5	2.3	3.9	2.7
Concrete	2180	9.15	1.8	1.5	2.2	1.7	1.5	1.0	1.7	1.1
Concrete	2180	36.6	1.9	1.7	2.4	1.8	1.6	1.2	1.9	1.2
Concrete	4920	9.15	1.8	2.2	2.4	2.5	1.9	1.7	2.2	1.7
Concrete	4920	36.6	1.9	2.3	2.4	2.9	2.0	1.8	2.6	2.0
Eytelwein Formula Safe Resistance										
Steel	291	9.15	1.4	0.6	1.8	0.7	1.0	0.4	1.4	0.4
Steel	291	36.6	1.3	0.7	1.5	0.7	0.8	0.3	0.9	0.3
Steel	2330	9.15	2.7	2.9	3.9	3.6	2.1	1.4	3.0	1.8
Steel	2330	36.6	5.4	9.0	8.5	11.8	3.1	3.5	4.9	4.4
Concrete	2180	9.15	1.8	1.4	2.2	1.6	1.3	0.7	1.0	0.6
Concrete	2180	36.6	3.6	4.7	4.3	4.9	2.1	1.8	2.4	1.8
Concrete	4920	9.15	2.6	3.5	3.2	4.1	1.9	1.7	2.2	1.7
Concrete	4920	36.6	7.1	12.7	8.6	16.0	4.2	5.3	5.4	5.9
Hiley Formula Ultimate Resistance										
Steel	291	9.15	0.8	1.0	0.9	1.1	0.8	0.9	0.9	0.9
Steel	291	36.6	1.1	1.3	1.0	1.0	1.1	1.2	0.9	0.9
Steel	2330	9.15	0.9	1.6	1.3	1.9	1.0	1.3	1.3	1.5
Steel	2330	36.6	1.2	2.2	1.8	2.6	1.3	1.9	1.7	2.0
Concrete	2180	9.15	1.1	1.5	1.3	1.6	1.0	1.2	0.8	1.1
Concrete	2180	36.6	1.9	2.7	2.1	2.5	1.7	2.2	1.7	1.8
Concrete	4920	9.15	1.3	2.5	1.6	2.6	1.3	1.9	1.4	1.8
Concrete	4920	36.6	2.0	3.6	2.2	4.3	2.1	3.1	2.5	3.1

7.5.5   Application to practical problems

Considerable efforts have been made to determine the reliability of the wave equation analysis when applied to practical piling problems and to determine whether it provides ultimate load predictions of increased accuracy as compared with the conventional driving formulae. Typical comparisons between predicted ultimate load versus set curves and the measured ultimate load from a load test are shown in *Figures 7.32a, b* and *c*. These pile load tests have been described by Golder (1953) and a number of predicted curves are shown for various values of the soil parameters. The load test point generally lies between the predicted curves and on the basis of such comparisons, 'best-fit' values of the soil parameters may be determined. Such a procedure

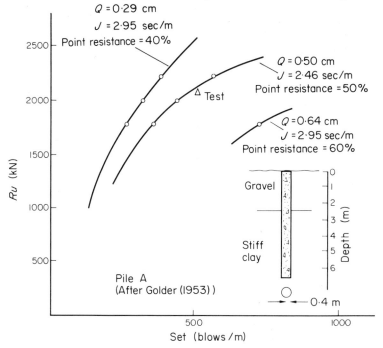

Figure 7.32a   Comparisons between predicted and observed pile resistance

has been used by Forehand and Reese (1964) to obtain the values of soil parameters in *Table 7.9*.

Hirsch *et al.* (1970) have reported a number of correlations of wave equation solutions with full scale tests to failure for piles in sand and in clay. For the piles in sand, damping constants of J(point) = 0.1 and J'(side) = J(point)/3 were found to give the best correlation. *Figure 7.33* shows the accuracy of the correlation for piles in sand to be about $\pm 25\%$. For the piles in clay, the damping constants J(point) = 0.3 and J'(side) = 0.1 gave the best correlation. *Figure 7.34* shows the accuracy of the correlation in this case to be approximately $\pm 50\%$.

A statistical analysis of 78 load tests was made by Sorensen and Hansen (1957) using the wave equation and various driving formulae to predict the

276

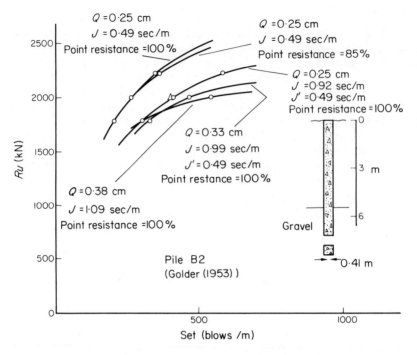

Figure 7.32b   Comparisons between predicted and observed pile resistance

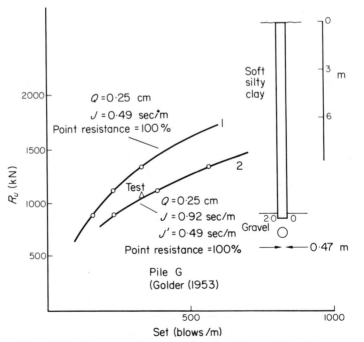

Figure 7.32c   Comparisons between predicted and observed pile resistance

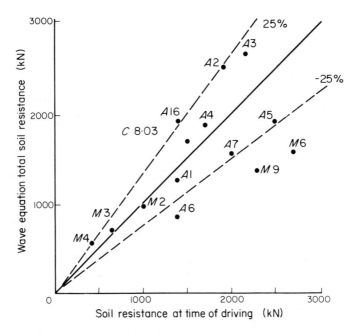

Figure 7.33   Comparison of wave-equation predicted soil resistance to soil resistance determined by load tests for piles driven in sands (after Hirsch et al. (1970))

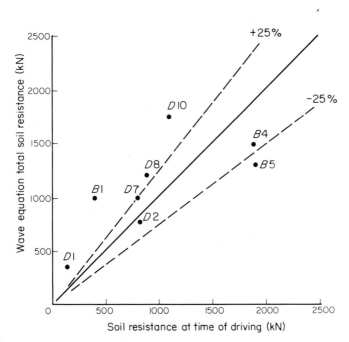

Figure 7.34   Comparison of wave-equation predicted soil resistance to soil resistance determined by load tests for piles driven in clay (after Hirsch et al. (1970))

ultimate load. They concluded that the wave equation was at least as good as the best of the pile driving formulae and derived the following data:

(1) The standard derivation on the ratio of predicted to observed ultimate load was 0.23.
(2) The upper limit for 96% safety, if the lower limit is 1.0, was 4.0.
(3) The nominal safety factor, i.e. the safety factor if only 2% of the predicted loads are permitted to have an actual safety factor less than 1.0, was 2.7.

A statistical analysis of the results reported by Hirsch *et al.* (1970) gave values consistent with those derived by Sorensen and Hansen.

In addition to its apparent success in predicting the load capacity, the wave equation approach also provides a rational means of evaluating the effects of significant parameters such as type and size of pile driving hammer, driving assemblies (cap block, helmet and cushion block), type and size of pile, and soil condition. From such an analysis, appropriate piles and driving equipment can be selected to avoid construction problems such as excessive driving stresses, pile breakage and inadequate equipment for the desired penetration or load capacity.

BIBLIOGRAPHY

BEREZANTZEV, V. G., KHRISTOFOROV, V., and GOLUBKOV, V. (1961). 'Load Bearing Capacity and Deformation of Piled Foundations', *Proc. 5th Int. Conf. SMFE*, **2**, 11–15.
BOWLES, J. E. (1968). *Foundation Analysis and Design*, McGraw-Hill, New York.
BURLAND, J. B., BUTLER, F. G., and DUNICAN, P. (1966). 'The Behaviour and Design of Large Diameter Bored Piles in Stiff Clay', *Proc. Symp. on Large Bored Piles*, *I.C.E.*, 51–71, London.
BUTTERFIELD, R., and BANERJEE, P. K. (1970). 'A Note on the Problem of a Pile-Reinforced Half Space', *Geotechnique*, **20**, 1, 100–103.
BUTTERFIELD, R., and BANERJEE, P. K. (1971). 'The Elastic Analysis of Compressible Piles and Pile Groups', *Geotechnique*, **21**, 43–60.
COYLE, H. M., and REESE, L. C. (1966). 'Load Transfer for Axially Loaded Piles in Clay', *Jour. Soil Mech. and Fdn. Eng. Div., ASCE*, **92**, SM 2, 1–26.
COYLE, H. M., and SULAIMAN, T. (1967). 'Skin Friction for Steel Piles in Sand', *Jour. Soil Mech. and Fdn. Eng. Div., ASCE*, **93**, SM 6.
D'APPOLONIA, E., and ROMUALDI, J. P. (1963). 'Load Transfer in End-Bearing Steel H-Piles', *Jour. Soil Mech. and Fdn. Eng. Div., ASCE*, **89**, SM 2, 1–25.
D'APPOLONIA, D. J., and LAMBE, T. W. (1971). 'Four Foundations on End-Bearing Piles', *Jour. Soil Mech. and Fdn. Eng. Div., ASCE*, **97**, SM 1, 59–75.
DAVIS, E. H., and POULOS, H. G. (1972). 'Analysis of Pile-Raft Systems', *Aust. Geomechs. Jnl.*, **G2**, 1, 21–27.
DAVISSON, M. T., and MCDONALD, V. J. (1969). 'Energy Measurements for a Diesel Hammer', *Performance of Deep Foundations, ASTM*, STP 444, 295–337.
ELLISON, R. D., D'APPOLONIA, E., and THIERS, G. R. (1971). 'Load-Deformation Mechanism for Bored Piles', *Jour. Soil Mech. and Fdn. Eng. Div., ASCE*, **97**, SM 4, 661–678.
FOREHAND, P. W., and REESE, J. L. (1964). 'Prediction of Pile Capacity by the Wave Equation', *Jour. Soil Mech. and Fdn. Eng. Div., ASCE*, **90**, SM 2, 1–25.
GLANVILLE, W. H., GRIME, G., FOX, E. N., and DAVIES, W. W. (1938). 'An Investigation of the Stresses in Reinforced Concrete Piles During Driving', *Br. Bldg. Res. Bd. Tech. Paper No. 20*, DSIR.
GOLDER, H. Q. (1953). 'Some Loading Tests to Failure on Piles', *Proc. 3rd Int. Conf. SMFE*, *Zürich*, **2**, 41–46.
HANNA, T. H. (1963). 'Model Studies of Foundation Groups in Sand', *Geotechnique*, **13**, 334–351.
HIRSCH, T. J., LOWERY, L. L., COYLE, H. M., and SAMSON, C. H. (1970). 'Pile-Driving Analysis by One-Dimensional Wave Theory: State of the Art', *High. Res. Record*, No. 333, 33–54.
HOUSEL, W. S. (1966). 'Pile Load Capacity: Estimates and Test Results', *Jour. Soil Mech. and Fdn. Eng. Div., ASCE*, **92**, SM 4, 1–30.

ISAACS, D. V. (1931). 'Reinforced Concrete Pile Formulae', *Trans. Instn. Engrs., Aust.*, **12**, 312–323.

KOIZUMI, Y., and ITO, K. (1967). 'Field Tests with Regard to Pile Driving and Bearing Capacity of Piled Foundations', *Soil and Fndn.*, **7**, 3, 30–53.

LOWERY, L. L., HIRSCH, T. J., EDWARDS, T. C., COYLE, H. M., and SAMSON, C. H. (1969). 'Use of the Wave Equation to Predict Soil Resistance on a Pile During Driving', *Spec. Session No. 8, 7th Int. Conf. SMFE, Mexico City.*

MANSUR, C. I., and KAUFMAN, R. I. (1956). 'Pile Tests, Low-Sill Structure, Old River La', *Jour. Soil Mech. and Fdn. Eng. Div., ASCE*, **82**, SM 4, Proc. Paper 1079.

MATTES, N. S. (1969). 'The Influence of Radial Displacement Compatability on Pile Settlements', *Geotechnique*, **19**, 157–159.

MATTES, N. S., and POULOS, H. G. (1969). 'Settlement of Single Compressible Pile', *Jour. Soil Mech. and Fdn. Eng. Div., ASCE*, **95**, SM 1, 189–207.

MATTES, N. S., and POULOS, H. G. (1971). 'Model Tests on Piles in Clay', *Proc. 1st Aust.–N.Z. Conf. on Geomechs.*, Melbourne, 254–259.

MATTES, N. S. (1972). 'The Analysis of Settlement of Piles and Pile Groups in Clay Soils', *Ph.D. Thesis*, University of Sydney.

MCCAMMON, N. R., and GOLDER, H. Q. (1970). 'Some Loading Tests on Long Pipe Piles', *Geotechnique*, **20**, 171–184.

MCCLELLAND, B., FOCHT, J. A., and EMRICH, W. J. (1969). 'Problems in the Design and Installation of Offshore Piles', *Jour. Soil Mech. and Fdn. Eng. Div., ASCE*, **95**, SM 6, 1491–1514.

MINDLIN, R. D. (1936). 'Force at a Point in the Interior of a Semi-Infinite Solid', *Physics*, **7**, 195.

MOHAN, D., JAIN, G. S., and KUMAR, V. (1963). 'Load Bearing Capacity of Piles', *Geotechnique*, **13**, 76.

MOSLEY, E. T., and RAAMOT, T. (1970). 'Pile-Driving Formulas', *High. Res. Record*, No. 333, 23–32.

POULOS, H. G., and DAVIS, E. H. (1968). 'The Settlement Behaviour of Axially-Loaded Incompressible Piles and Piers', *Geotechnique*, **18**, 351–371.

POULOS, H. G. (1968). 'Analysis of the Settlement of Pile Groups', *Geotechnique*, **18**, 449–471.

POULOS, H. G., and MATTES, N. S. (1969). 'The Behaviour of Axially Loaded End-Bearing Piles', *Geotechnique*, **19**, 285–300.

POULOS, H. G., and MATTES, N. S. (1971a). 'Settlement and Load Distribution Analysis of Pile Groups', *Aust. Geomechs. Jnl.*, **G1** No. 1, 18–28.

POULOS, H. G., and MATTES, N. S. (1971b). 'Displacements in a Soil Mass Due to Pile Groups', *Aust. Geomechs. Jnl.*, **G1** No. 1, 29–35.

POULOS, H. G. (1972a). 'Load-Settlement Prediction of Piles and Piers', *Jour. Soil Mech. and Fdn. Eng. Div., ASCE*, **98**, SM 9, 379–397.

POULOS, H. G. (1972b). 'Settlement Analysis of Two Buildings on End-Bearing Piles', *Proc. 3rd South-East Asian Conf. Soil Eng.*, Hong Kong.

SAFFEREY, M. R., and TATE, A. P. K. (1961). 'Model Tests on Pile Groups in a Clay Soil with Particular Reference to the Behaviour of the Group when it is Loaded Eccentrically', *Proc. 5th Int. Conf. SMFE, Paris*, **2**, 129–134.

SALAS, J. A. J., and BELZUNCE, J. A. (1965). 'Resolution Theorique de la Distribution des Forces dans les Pieux', *Proc. 6th Int. Conf. SMFE, Montreal*, **2**, 309–313.

SAMSON, C. H., HIRSCH, T. J., and LOWERY, L. L. (1963). 'Computer Study of Dynamic Behaviour of Piling', *Jour. Struct. Div., ASCE*, **89**, ST 4, 413–449.

SEED, H. B., and REESE, L. C. (1957). 'The Action of Soft Clay Along Friction Piles', *Trans., ASCE*, **122**, 731–754.

SMITH, E. A. L. (1960). 'Pile Driving Analysis by the Wave Equation', *Jour. Soil Mech. and Fdn. Eng. Div., ASCE*, **86**, SM 4, 35–61.

SORENSEN, T., and HANSEN, B. (1957). Pile Driving Formulae—An Investigation Based on Dimensional Considerations and a Statistical Analysis', *Proc. 4th Int. Conf. SMFE*, **2**, 61–65.

SOWERS, G. F., MARTIN, C. B., WILSON, L. L., and FAUSOLD, M. (1961). 'The Bearing Capacity of Friction Pile Groups in Homogeneous Clay from Model Studies', *Proc. 5th Int. Conf. SMFE, Paris*, **2**, 155–159.

THURMAN, A. G., and D'APPOLONIA, E. (1965). 'Computed Movement of Friction and End-Bearing Piles Embedded in Uniform and Stratified Soils', *Proc. 6th Int. Conf. SMFE, Montreal*, **2**, 323–327.

WHITAKER, T. (1957). 'Experiments with Model Piles in Groups', *Geotechnique*, **7**, 147–167.

WHITAKER, T., and COOKE, R. W. (1966). 'An Investigation of the Shaft and Base Resistances of Large Bored Piles in London Clay', *Proc. Symp. on Large Bored Piles, I.C.E.*, 7–49. London.

WILLIAMS, G. M. J., and COLMAN, R. B. (1965). 'The Design of Piles and Cylinder Foundations in Stiff, Fissured Clay'. *Proc. 6th Int. Conf. SMFE, Montreal*, **2**, 347–351.

Index